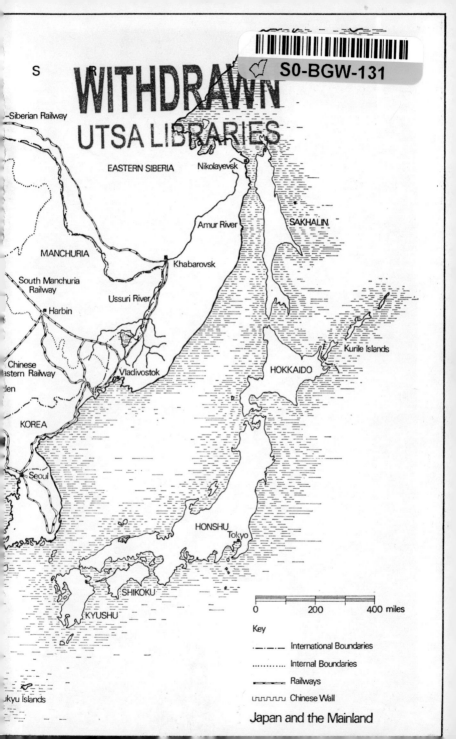

C01177601

DS
835
.C66
1970

B219443

JAPAN: An Historical and Cultural Introduction

JAPAN
An Historical and Cultural Introduction

E. M. Cooper

 Pergamon Press Australia

Pergamon Press (Australia) Pty Limited, 19a Boundary Street,
  Rushcutters Bay, NSW 2011
Pergamon Press Ltd, Headington Hill Hall, Oxford OX3 OBW
Pergamon Press Inc., Fairview Park, Elmsford, NY 10523, USA
Pergamon of Canada Ltd, 207 Queen's Quay West, Toronto 1
Friedr. Vieweg & Sohn, GmbH, Postfach 185, 3300 Braunschweig,
  West Germany

© 1970 E. M. Cooper
First published in Australia 1970 by
Pergamon Press (Australia) Pty Limited
Printed in Australia by The Specialty Press Limited, Melbourne
Registered in Australia for transmission by post as a book.
National Library of Australia Card Numbers and SBNs
08 087184 1 (flexi)
08 017539 2 (hard)

This book is copyright. No part of it may be reproduced,
stored in a retrieval system, or transmitted, in any form
or by any means, electronic, mechanical, photocopying,
recording or otherwise, without prior permission of Pergamon
Press (Australia) Pty Limited.

Contents

List of Illustrations, vii
Acknowledgements, viii
1 Geographic Setting, *1*
2 Early History, *7*
3 The Japanese Language, *15*
4 The Rise of Feudalism, *20*
5 The Ashikaga Shogunate, *26*
6 Religion, *30*
7 The End of the Civil War, *34*
8 European Interest in the East, *39*
9 The Seclusion Period, *45*
10 American Interest in Japan, *51*
11 The Fall of the Tokugawa Shogunate, *54*
12 Japanese Drama, *58*
13 The Meiji Restoration, *63*
14 The Meiji Constitution, *70*
15 Foreign Affairs, 1868-1905, *76*
16 Relations with America, *84*
17 Japan as a Major World Power, *87*

18 Japanese Politics, 1918-1945, *93*
19 Foreign Affairs, 1914-1939, *98*
20 The War in the Pacific, *107*
21 Foreign Occupation, *115*
22 Japan After the War, *120*
23 Conclusion, *125*

Appendices, *128*
Appendix I: The Forty-Seven Ronin, 128
Appendix II: Will Adams, 132
Appendix III: Soka Gakkai, 134

Glossary of Geographical Terms, *136*

Further Reading, *138*

Questions and Suggestions for Further Work, *140*

Time Chart, *145*

Index, *149*

List of Illustrations

Maps: Japan, *front endpaper*
 Japan in relation to the mainland, *back endpaper*

Ho-o-do (Phoenix Hall) of the Byodoin Temple, Nara, *56*
Great Buddha Statue of Todaiji Temple, Nara, *56*
Hirosaki Castle at cherry blossom time, *57*
Doll in costume from the Noh play *Hagoromo* or
 The Robe of Feathers, *72*
Kabuki warrior portrayed by actor Mitsugoro Bando, *72*
Kakizome—the ceremony of 'the first writing of the New Year', *73*
Dress of the period of the Meiji Restoration, *73*
Snow Festival at Sapporo City, Hokkaido, *104*
Meiji Shrine, built in pure Shinto style, *105*
The Japanese Imperial Family, *120*
Wedding of Crown Prince Akihito and
 Crown Princess Michiko, *120*
New Year's Day begins with a toast, *121*
Family watching television—a typical evening scene in
 modern Japan, *121*

Page numbers indicate pages facing illustrations

Acknowledgements

The author and the publishers wish to thank the following people for their assistance: Miss Kuno, Japanese Vice-Consul in Melbourne; Miss Sugizaki of the Japanese Consulate, Sydney; Mr Toshio Kutsuno, Mr Kazuo Kubo, Mr V. N. Jernakov, Mr and Mrs M. Capon, Mr Don Mussared, Miss Kathy Cooper, Mr Ponting and the staff of the Yallourn Library.

The endpaper maps are the work of Miss Lesley Mansfield; the charming drawing is by Miss Shigeko Yoshidomi of Fukawa Junior High School, Yamaguchi. The photographs are by courtesy of the Japanese Consulates in Melbourne and Sydney.

I
Geographic Setting

Japan consists of four main islands—Hokkaido (or Yezo), Honshu, Kyushu and Shikoku—and over three thousand minor islands.

Its position is so similar in relation to the continent of Asia to England's to the continent of Europe, that historians are often tempted into drawing comparisons. Actually, the similarity ends there. While England was for the first millenium after the birth of Christ subjected to a series of invasions and occupations by continental forces, Japan has never until recently been affected by outside influences through conquest. She has imbibed Chinese culture voluntarily and not until the present day did conquest force on her an alien culture—this time American.

The Build of Japan
The country is mountainous, with many volcanoes, some extinct, others still active. Earth tremors occur daily in some part or other of the islands and earthquakes are comparatively frequent. Hot springs and steam clouds are a constant reminder of the heat that lies below the surface, especially in Hokkaido.

Only about a fifth of the land can be cultivated, often in fields terraced out of the hillsides. Farms are very small, but intensively cultivated.

In Honshu farms may be only two or three acres. In Hokkaido they are generally bigger, perhaps fifty acres, and farmers here are more prosperous than in the southern islands.

Plains occur around Kyoto and Tokyo (the Kanto) and there is an extensive plain near Sapporo in Hokkaido. Although the average height above sea level is not great, except in north Hokkaido, it is worth noting that the famous Mount Fuji is over twelve thousand feet high.

Communications

The Japanese are great travellers in their own country. They are very conscious of its beauty and proud of its history. A prominent feature of the education system is the encouragement of such feelings by school visits to places of interest.

Besides the roads, which vary from up-to-date highways with elaborate fly-over systems to rutted tracks in remoter areas, there is a good railway complex run by the state. This is supplemented by private railways. The Japanese are justly proud of their super-express, which runs between Tokyo and Osaka, averaging 120 miles an hour.

Products

Both light and heavy industry is well developed. Japan ranks third after the United States and the U.S.S.R. as a steel producer. She has long led the world in shipbuilding and now ranks third also as a producer of motor vehicles. Given the raw materials Japan can and will produce anything. She has long since lived down the reputation for shoddy workmanship that she acquired in her first burst of industrialisation after the First World War.

Japan is deficient in raw materials, and industry has to rely heavily on imports. The country has deposits of coal, copper, chromium, gold, silver and zinc, but only in small quantities. Three-quarters of Japan's iron ore needs have to be imported. Coal is mined in Hokkaido, in Honshu north of Tokyo and in north Kyushu, but coking coal has to be imported. Sulphur forms the basis for the manufacture of paper, celluloid and rayon. Salt has to be obtained from sea water and large imports are required by industry.

In agriculture, rice is the chief product. Wheat, rye, barley and oats are grown extensively in the colder Hokkaido. Tea, silk, flax, hemp, pyrethrum and tobacco are other products. Fruit growing is increasing, as is dairying. Hokkaido leads the way in breaking away from the traditional dependency on rice culture, but the Japanese farmer has shown a readiness to extend his range of crops and try new methods. This is unusual for peasant farmers, traditionally a very conservative group.

Farming in Japan relies heavily on fertilisers as the soil is poor. Artificial fertilisers are beginning to replace the reliance on human excrement that used to be a feature of the Japanese countryside.

Fishing is an important industry and canned fish an important export. Cultured pearls are an interesting result of Japanese ingenuity and proximity to the sea.

Geographic and Ethnic Factors in Japan's Progress

The Japanese people have outstanding qualities of perseverance and application. They are willing and eager to learn. They are not hampered by religious beliefs or social taboos such as restrict the development of some other Asian countries. Both government and people realise the value of education, so that Japan is among the most literate countries, not merely in Asia, but in the world.

The position of the Japanese islands in relation to the continent has meant that cultural exchange and trade with China was never completely cut off, even during the seclusion period of Japan's history. Now Japan commands the approach to the mainland, with obvious trading benefits.

On the other hand, the sea has protected Japan from invasion until comparatively recently. Her island make-up produced a sea-going people. In fact, from a study of geography it would be reasonable to assume that Japan should have been one of the great trading and colonising nations of the world. Other factors, however, contributed to strengthen the policy of isolation followed by her rulers for two hundred years. For example, climatic and topographic features made the east coast of the country the most highly developed area. It looks towards the ocean rather than to the mainland. The capital cities were isolated by nature, by mountains to the west and ocean to the east.

Climatically Japan lies in the temperate zone. It is well watered and not subject to the periodic droughts that afflict China and India. The distinctive style of architecture can be ascribed in part to the climate (it also must be influenced by the availability of the building materials and the fact that light structures are safer in earthquakes than heavier ones would be). Some writers argue that Japanese buildings are very inadequate in the colder weather and consider the style must have been brought to Japan by the ancestors of the Japanese, coming from a warmer climate. Certainly the people are very conservative, and although their building materials are perishable wood and paper, the old buildings are usually rebuilt carefully so that the new is a faithful copy of the old. Hokkaido is of course much cooler generally than the other islands, with heavy snowfalls in winter. Buildings are more substantial and stone is in common use.

The natural beauty of Japanese scenery has developed the artistic abilities of the people. Many of the country's products owe their special appeal to their designers' aesthetic sense.

The division of the country by mountains and sea led to the long maintenance of feudalism in Japan: a factor of obvious significance in her subsequent history.

Finally, geographical factors have influenced the diet of the Japanese people. The main foods have for centuries been rice and fish. Until recently this restricted diet contributed to the small size of the people and deficiency in vitamins was blamed for instances of poor eyesight. Now that a more varied fare, with meat and milk products, is possible and is gaining favour, a notable change is taking place in the build of the younger generation.

Japanese Cities

Tokyo has been the capital of Japan since 1868. It has a population of more than eleven million.

Osaka is the second largest city in Japan, and is an important commercial and industrial centre. It is situated at the mouth of the river Yodo. A fine network of canals interlaces its busy streets. The chief products are cotton goods, synthetic textiles and steel.

Yokohama is a busy port catering for shipping from all over the world.

Kobe, a manufacturing city and port, is mainly concerned with shipbuilding and heavy industry.

Hiroshima is an important port in southwest Honshu.

Nagasaki, in the west of Kyushu, has coal, steel, cement, glass and chemical manufacturing.

Sapporo is the largest city and administrative, educational and commercial centre of Hokkaido.

Geographic Setting 5

Kyoto, Nara, Nikko and Kamakura are well known to tourists for their historical and artistic treasures.

Island Territories Occupied by Japan up to 1945
Sakhalin (or Karafuto): The southern part of this large and strategically placed island was acquired from Russia in 1905. After the 1917 revolution in Russia, Japanese troops moved into North Sakhalin, but withdrew in 1925 in return for oil concessions. The oil Japan so bady needed had been discovered in the north of the island. Always a subject for dispute, the island became Russian in 1945.

The Kurile Islands: Hokkaido is really an administrative term that embraces the northern island of the main group, Yezo, and the Kurile islands. After 1945 the latter were occupied by Russia. Yezo and many of the outlying islands were not fully under the control of the mainland government until the nineteenth century.

The Bonin Islands: These islands to the south east of Japan are less than fifty square miles in area. After 1945 they were administered by the United States of America, but were returned to Japan in 1968.

The Ryukyu Islands: The people are closely related to the Japanese, but they had their own ruler and paid tribute to China until the 1870's, when the Japanese established their authority. In 1945 the Americans took them over because of their strategic position, but the Japanese are not dispossessed. The important American military and air base, Okinawa, is the largest of the Ryukyus.

Taiwan (Formosa): This large island was annexed by the Japanese after the war with China in 1894-95. The majority of the population is of Chinese stock, whose ancestors emigrated from the neighbouring mainland coast and pushed the indigenous people back into the hills, where they still live, many in a very backward state. They appear to be akin to the Malays. Until recent years they were feared for their fierceness and sea travellers in the area trembled at the thought of being cast up on the shores of Taiwan.

Taiwan was returned to China in 1945 and in 1949 became the home of the Chiang Kai-shek forces after they were pushed from the mainland by the Communist Chinese. The island is rich agriculturally and has a little coal, oil and useful timber. It is being developed with the help of American capital.

Japanese Mainland Possessions Lost after 1945

Manchuria: To the north east of China proper, Manchuria is a vast plain bounded by mountains. The Liaotung Peninsula dominates the sea approach to north China. Relatively thinly populated until recent times, Manchuria has for a long time been the object of Russian and Japanese ambition. In the 1930's, turmoil in China left the way open for Japan to establish control, though in theory Manchuria was an independent state. It has coal and oil shale and a good communication system, thanks to Russian and Japanese interest. In 1945 the Japanese were forced to withdraw and the Communist government of China now has firm control over the area. Migration from the northern states of China has been encouraged to build up the population.

Korea: This peninsula has been a continual battleground, subject to invasion from China and Japan. A glance at the map on the front endpaper of this book will show that neither China, Japan nor Russia can be happy to see Korea under the control of a potential enemy. After the defeat of Russia by Japan in 1905 the peninsula came under Japan's control, and it was officially annexed in 1910. After 1945 it became independent again, but is now divided into two parts. Apart from its strategic value, Korea has little to offer the invader; it is very mountainous, and the deposits of coal and iron are small. However, the reafforestation schemes initiated by the Japanese are beginning to produce vast stands of timber.

Some parts of China were occupied by Japanese troops in the 1930's. From their foothold in Manchuria they were beginning to push into Inner Mongolia, and they had ambitious plans for the lands to the south of their puppet state. After the war of 1941-45, however, the Japanese were forced to give up all possessions on the mainland.

2
Early History

The modern Japanese appear to be a mixture of three strains—South Chinese, Malay and Ainu. Scientific confirmation of this theory has yet to be made, but there seem to be grounds for arguing that there were two main invasions from the mainland of Chinese-Malay type people. The newcomers found the islands already occupied by the Ainu. The early inhabitants were gradually pushed back by the invaders, until now they are found only in Hokkaido.

The Ainu are markedly different from the Japanese, as their eyes lack the Mongoloid fold that gives the distinctive appearance of almond-shape to the eyes of their neighbours. Often known as 'the hairy Ainu', the men sport long beards and shaggy hair. Some of the women tattoo blue beards on their chins.

Culturally these people have affinities with tribes of northern Russia, such as the Samoyeds and Buryats. They believe in a multitude of demons and nature spirits, which can be controlled by the Shaman or witch doctor. The bear has a prominent place in their ritual. Baby bears are brought up carefully and fed lavishly merely so that they can be ceremonially slaughtered. The Ainu claim that the bear skin imprisons a spirit which is glad to be set free.

In modern times the Ainu have become a tourist attraction, with the result that there is nothing to be seen but a parody of their former customs. They are gradually losing their characteristics as a separate race because of a declining birthrate and intermarriage with the Japanese.

Myths and Legends
Early Japanese history is largely a matter of legend and it is not known for how long the stories have been passed on by word

of mouth before writing was introduced from China. Traditionally the first emperor was Jimmu Tenno, whose date is given as 660 BC. The present emperor claims direct descent from him, and through Jimmu, from the Sun God.

Bearing in mind the climatic conditions, volcanic eruptions and so on, which must have given the Japanese islands their present form, it is conceivable that the myths of early Japan are very ancient indeed. The gods Izanami and Izanagi are supposed to have stirred up the sea with a lance. When they withdrew it, a drop of water fell from the tip and congealed into an island. One of their children, the Fire God, burned his mother to death, and she was buried at Izumo on the south west coast of Japan, facing the mainland (another Izumo in northwest Honshu also claims this honour). Recent excavations have unearthed neolithic (New Stone Age) settlements in this area with stone implements and hand-moulded pottery, tentatively dated as far back as 3000 BC. In the third century BC, bronze and iron were introduced from the mainland.

To return to the myths: the father Izanagi sought for his lost spouse in the underworld, but in vain. On his return he had to purify himself by washing and in doing so produced the Sun Goddess, Amaterasu, the Moon Goddess and the Storm God.

The Sun Goddess and the Storm God fought and quarrelled continually, but eventually the goddess prevailed. The god fled to Korea and returned in a chastened mood to settle in Izumo. Amaterasu sent her grandson with instructions to make himself the ruler of the 'Central Land of Reed Plains' (Japan, or more specifically, Honshu). He is supposed to have landed on Mount Takachiho in Kyushu. From there he began his progress of conquest northwards. Jimmu Tenno, who is reputed to have established himself successfully at last in Yamato in central Japan, was his great-grandson.

The name Tenno, meaning 'Son of Heaven', has been applied posthumously to the names of all Japanese emperors ever since. Though tradition ascribes Jimmu's accession to the throne as 660 BC, modern historians consider 60 BC more likely, so that Jimmu Tenno is roughly a contemporary of Julius Caesar.

This is of course only a brief summary of some of the stories. How the imperial regalia came to include the Sacred Mirror,

Jewels and Sword, is also elaborately described in the legends.

The story-tellers evidently kept alive a tradition of successive waves of migration from the mainland, passed down by word of mouth, long before Japanese history came to be written down. One group seems to have settled round Izumo, while the other pushed on from Kyushu to become the more powerful and to establish eventual control over the islands.

It is impossible to know how many of the modern Japanese still believe their emperor is descended from the Sun Goddess but undoubtedly he holds a special place in the eyes of his subjects.

The Beginning of Civilisation

Archaeologists may one day confirm the basic truth of the Japanese myths, as they have already proved there is a groundwork of fact in stories such as the Minotaur legend in Crete. The earliest prehistoric pottery (Proto-Jomon) is tentatively dated about 3000 BC. It has been found only in Hokkaido and Honshu. From this and other evidence it is assumed that the culture originated in Siberia.

The name Jomon means 'straw-rope pattern'. The decoration characteristic of the Japanese Stone Age pottery is achieved by pressing rope or a rope-bound stick on to the damp clay.

Until about 300 BC the Jomon culture continued to flourish. Excavations show a gradual development. The pottery improves in shape and quality; remains of houses appear, at first singly, then grouped into villages. Stone axes are supplemented by wooden bows, similar to those still used by the Ainu. Shell ornaments and lacquered wood show that towards the end of the Jomon period some objects were prized for their beauty, not merely for their utility.

Bones of domesticated dogs and horses have been found in excavation sites in north Japan, but so far nothing that would indicate a knowledge of agriculture.

From about 300 BC a new culture pattern emerges. This culture is known as Yayoi, from the area of Tokyo in which extensive remains have been found. The people who practised the culture must have entered the country from the south, but archaeologists differ as to whether they came from China, Indonesia or the Philippines. They were iron-using and practised agriculture. For the first time rice was introduced into Japan. These people appear to have established themselves in Kyushu and then pushed on into Honshu to settle in the Tokyo plain. Here was the centre of the state known as Yamato.

Archaeologists have learned that the state centred around Yamato was iron-using. The people had begun to cultivate rice. They had simple houses, made cloth and pottery, and were very fond of jewellery. Their religion was a simplified version of modern Shinto. Everything had its 'Kami' or spirit, which must be propitiated by offerings and prayers. The emperor was at once the chief priest and God himself with, as we have seen, a special relationship to the sun.

Early Japan appears to have been organised into clan units with a family kinship basis and worshipping individual gods, even claiming descent from a particular god. The family of the leading clan gradually achieved pre-eminence and its leader acquired the title we translate as emperor.

From the beginning Japanese society was divided into classes. After the aristocracy, with their supernatural relations, came the inferior breeds of men. The artisans were organised into trade guilds, and fared not too badly, but under them were the slaves with no rights at all. Among these would have been the survivors of the early inhabitants.

Some idea of the customs can be gained from the accounts of the rulers of this early period. As they were not written down until the eighth century AD, they can hardly be accorded the authority of true history.

The Emperor Suinin is remembered for having stopped the practice of burying the personal attendants of the dead at the funeral of important people. The servants were thought to administer to their master's wants after death as before, but by the time of Suinin it was felt that clay models of the servants were just as good. Like the models found in Egyptian tombs, these Japanese ones are studied with interest for the light they throw on the period. Known as *haniwa*, the figures usually show warriors with armour, swords and horses. The dress is quite different from later Japanese dress. The men wear baggy trousers and tight-fitting jackets, while the women wear a blouse and skirt.

Early History

The Empress Jingo is an elusive figure, whose existence is not confirmed by contemporary Chinese records, as one would expect if her reputation is correct. She is supposed to have conquered Korea. At this time Korea was divided into three separate kingdoms continually at war with each other and inviting alliance from Japan. True or not, the story of Jingo's conquest and of the promise of everlasting allegiance to Japan claimed to have been given by Korea, has been the basis of Japan's interest in that country ever since.

In view of the later very subservient position held by women, it is interesting to note the prominence of their position in these early times. They were often rulers, and in the Shinto religion they functioned as priestesses, a position they can still hold.

In 536 AD the system, so typical of Japan, by which an aristocratic family governed the country while the emperor became a mere figurehead, made its first appearance. The Soga family were able to attain this power by playing its rivals off against each other. The Sogas favoured the introduction of Buddhism from China because it gave them the opportunity to undermine the authority of those noble families who owed their influence to hereditary offices in the Shinto hierarchy.

The Impact of Chinese Culture

From the end of the sixth century AD Japan began to be increasingly affected by her proximity to China. The history of her great mainland neighbour stretches back long before the birth of Christ and the Japanese were eager students of religion, art and government.

In the sixth century AD one of the kings of Korea sent a statue of Buddha, Buddhist writings and monks to the Japanese court. Apart from the Soga clan, the imperial prince Shotoku was attracted to the new philosophy and his favour gave the initial impetus to the spread of Buddhism. Temples and monasteries were built, different indeed from the simple Shinto shrines. They became civilising centres, spreading Chinese culture. Like European monasteries, Buddhist institutions made themselves responsible for the provision of orphanages, infirmaries and schools. Buddhist scholars moved from one monastery to another and from Japan to China. Even the humblest Japanese could learn something of the new religion from the pictures of the life of Buddha that decorated the walls of temples or formed the design for silken embroidery. The shrines continued to attract their devotees and the two religions existed side by side and eventually mingled, until only experts could distinguish which practices were Buddhist and which Shinto.

At first Chinese culture was available to the Japanese only through contact with Korea, but soon they were sending their own envoys to the court of China. One record describes the 'hairy barbarians' (the Ainu) that they brought with them. Merchants and scholars came in growing numbers. These learned the Chinese language and in the fifth century AD began the adaptation of Chinese writing to Japanese speech (see Chapter 3 on Language). The distinctive form of painting and some architectural forms are derived from China.

In 710 AD a capital city, Nara, was laid out on Chinese lines—a copy of the Chinese capital so admired by Japanese visitors. Some of the buildings can still be seen as they were twelve hundred years ago. From 710 to 794 AD is known as the Nara Period in Japanese art. The most popular theme with artists was the life of Buddha, and the influence of Buddha's Indian origin can be seen in works of this period. The first Japanese books, however, were the *Kojiki* and the *Nihon Shoki*, dealing with the legendary beginnings of the nation. A third book of the Nara period was the *Manyoshu*, a book of poetry.

In 645 AD the Sogas were overthrown by the Fujiwaras. This aristocratic family had long resented the power of the Sogas. As hereditary chief priests of the Shinto religion, they naturally had suffered from the rival clan's support of Buddhism. The Sogas had become arrogant and overbearing. Their enemies at court were just awaiting a leader and an opportunity. The leader appeared in the shape of Prince Naka, son of Prince Shotoku. The latter's interest in Buddhism was purely philosophical, not political. As Regent he had chafed under Soga insolence, but his son was also a man of action. By a ruse, Iruka, the head of the Soga clan, was induced to appear at court without his sword. Before the throne itself he was hacked to death by Prince Naka, who then became emperor with the support of the Fujiwaras.

Early History

From that time onwards the custom of the emperor marrying a Fujiwara, as did Naka, has been almost unbroken. The present emperor showed uncommon resolution in choosing as his wife a princess from another family, and his children have even married commoners. The present empress went in fear of her life when it became known that the emperor's choice had fallen on her.

The Fujiwara are still Japan's second family, next only to that of the descendants of the Sun Goddess. Their close relationship with the royal family put them in a position of control, varying with the personalities of the emperors and the heads of the Fujiwara clan, for 500 years.

Many of the emperors were glad to abdicate at an early age and leave the burden of ruling to a Regent. Naturally a Fujiwara uncle was always available to take care of the business of government during the minority of the new emperor. When the boy attained his majority he usually was glad to have his uncle continue in power. A new title, Kampaku, or Dictator, was invented for the changed situation.

In 794 AD the capital was moved to Kyoto. Nara had fallen too much under the control of the Buddhist priesthood. The emperors were now more interested in living a life of luxury than following the dictates of religion. They set up a university in Kyoto and the town became the centre for the study of Chinese literature and art. The period is known as the Heian Period from the old name for Kyoto, Heian-kyo.

About 1000 AD the Lady Murasaki Shikibu wrote the first novel in any language, the *Tale of Genji*. It tells the amorous life story of Prince Genji, and recreates for us the atmosphere of the court at Kyoto, sophisticated and pleasure loving.

As soon as they came to power the Fujiwaras tried to introduce the Chinese system of government. In 646 AD the Taika (or Taikwa), the Great Reform, was introduced. Instead of government positions being in the hands of the aristocracy, the Chinese system of promotion by examination was introduced. The land was to be divided into administrative divisions and controlled by the examination-chosen bureaucracy. A new system of taxation was to be administered through the bureaucratic framework.

A very elaborate scheme for preventing the accumulation of

land in the hands of a few owners laid down that agricultural holdings were to be governed by the size of the family they were to support. Provision was made for redistribution where necessary.

If it had succeeded, the Great Reform would have centralised power in the hands of the emperor, or rather, the Fujiwara; but the clan structure and the geographic divisions of the country made this impossible. Offices were soon back in control of the aristocrats and the idea that class, not merit, should form the basis of promotion was more firmly established than ever. The redistribution of land proved too unwieldy from the start, while the taxation reform suffered the same fate as the others.

The Taiho Laws of 702 AD were another attempt to reorganise the government on Chinese lines and to introduce the Chinese legal system. The reform of the government was as fruitless as before, but the new code of laws, based on the Confucian system of strict obedience to authority, had lasting effect on Japan.

The Chinese family rule, that the father is the undisputed head and has to be obeyed unquestionably by all, was introduced. Gradually women were pushed into a position of definite subordination to men. The pattern of Japanese womanhood, where the wife's sole duty was to minister to the comforts of her husband and care for his home and family, asking nothing for herself, was set for the next twelve hundred years.

In the same way the vassal owed allegiance to his lord, while the priestly and divine nature of the emperor was emphasised. The basis for the establishment of the feudal system was laid down.

3
The Japanese Language

About the fifth century AD Japanese scholars who had visited China and learned the Chinese system of writing began to try to apply it to their own language. The Koreans also tried to adopt the elaborate symbols, but soon realised the difficulties and invented an alphabetic script for themselves. The Japanese, however, persisted in their imported system.

Chinese is a monosyllabic language. Obviously the number of monosyllabic variations is limited when only vowels and consonants are employed, but Chinese increases the variations by means of tone changes. This gives Chinese the peculiar sing-song effect typical of Asian languages of the same group, such as Siamese. A word spoken on high note is different in meaning from the same combination of consonants and vowels spoken in a falling or rising tone. For instance, the word for 'buy' and the word for 'sell' are both written *mai* in their anglicised form. In practice the different tones used in pronunciation prevent any confusion. In writing, of course, the distinction has to be made visual and this is done by using a different symbol for each word.

It is difficult to see how the Chinese could satisfactorily employ any substitute for their distinctive form of writing, though the anglicised forms are used and the Communist government has made valiant efforts to simplify the signs in an attempt to increase the literacy of the people. However, there is one big advantage: while there are many different spoken dialects in China, the written form can be understood by everyone. In fact, it could be read off into English by someone who had no knowledge of spoken Chinese.

In structure Chinese is rather a simple language. Its verbs do not change for person or tense. The subject precedes the verb and the object follows it. There are no plurals or genders, no adjective agreements. Where other languages have a wealth of synonyms often derived from the tongues of invaders, Chinese maintains a simple vocabulary and makes extensive use of combinations of basic ideas. For instance, 'famous' is 'have name'.

Japanese, on the other hand, has an extensive vocabulary. It is polysyllabic and its structure grammatically is quite different from Chinese. Subject and object are indicated by a post-position or particles. *Ga* or *wa* indicate that the noun is the subject. *Wo* is the mark of the object. The verb goes at the end of the sentence. Post-positions (placed *after* the word they modify) take the place of English prepositions (placed before). The verbs have endings which vary for tense and mood, though not for persons.

In order to cope with the Japanese grammar, scholars had to devise a system of symbols to represent the features of the language not present in Chinese. They came up with a syllabary of fifty symbols, representing five vowel sounds and their combinations with nine consonants. A further twenty-five combinations are obtained by addition of sound change indicators: a small mark changes 'k' to 'g'.

The whole of Japanese can be written satisfactorily in the syllabary *(hiragana);* in fact, the early literary works such as *Genji Monogatari*, written in the first quarter of the eleventh century, were entirely in this script. Children in present-day Japan learn hiragana first and only gradually are they expected to memorise the Chinese symbols *(kanji)*. In the first year of primary school they learn forty-six kanji and the fifty hiragana symbols. By the time they finish primary school in Grade Six they have memorised 881 kanji!

Not content with this, the Japanese have invented a further syllabic script known as *katakana*, which is used in writing foreign words. The symbols represent exactly the same syllables as hiragana and are equally as unsuitable for representing groups of consonants as found in European languages. For example 'table' is written *teburo*, although the pronunciation is little different from English. One can only conclude that katakana is an example of Japanese chauvinism.

After the Second World War the education system was reformed on American lines and an attempt was made to persuade the Japanese to write in Western script. Gradually the old system has been reintroduced and anyone claiming to be literate must master the kanji symbols—more or less of them according to his needs or his ability. However, there are obvious drawbacks. Much time must be spent in school in practising writing and a heavy burden is placed on the children. The Japanese typewriter is a formidable affair. The typist and the printer have to manipulate hundreds of symbols instead of an alphabet. It is virtually impossible to construct a dictionary for kanji. If the reader cannot guess the meaning from the context, if he has forgotten the symbol, his only hope is to find someone who can. Despite this, Japan has one of the highest literacy rates in the world!

The Japanese language also has many homonyms (words with the same spelling and sound). For instance, *akai hana* could mean 'a red flower' or 'a red nose'! In writing, a flower has one symbol and a nose another, so no confusion can arise. There is little difference made in pronunciation, except in tone that needs a quick ear to detect. The context usually makes the meaning quite clear. Exactly the same difficulty could be encountered in English—'a bright spring' could have three meanings. Japanese jokes, like English, often depend on this feature of the language.

There are two systems of numerals in Japanese. One, the original indigenous system, only goes up to ten. The Chinese system is used from ten onwards.

Some Asian languages make extensive use of numeral classifiers. In English we say 'two pairs of shoes', 'three sheets of paper' and so on. In Japanese a classifier is necessary with all nouns when used with the numbers up to ten, counting in the Chinese system. Often the numeral and the classifier combine in sound to produce what seems to be an entirely different word. For example, 'one dog' would be *ippiki (ichi + hiki)*, while 'one cupful' is *ippai (ichi + hai)*.

Novelists often transcribe Japanese dialogue into English in a manner which sounds ridiculous to us. A liberal scattering of 'honourables' and 'humbles' is considered necessary. Japanese use different nouns, pronouns and even verbs, when speaking to different classes of people. Although this is known as respect language and is used to indicate relations between members of the family, persons of different age and sex, or subordinates and persons in charge, it also serves to supplement the personal pronouns, which are not used much in Japanese. For instance, *tegami* would mean 'my letter', while *o-tegami* would be 'your letter'. Formerly, when the language was imperfectly understood this would have been translated 'your honourable letter'.

It is true of course that the Japanese do show what we would consider to be exaggerated respect for superiors, either in age or position, and the idea is carried on into the speech. The distinction is made in other languages too. Consider the difference between 'Dad' and 'Father'.

In giving an order we would temper the tone of our command according to the person being addressed. There is a big difference between 'Do it!', 'Please do it' and 'Would you mind doing this?'

Dates in Japanese will be stated in the Western system for the benefit of the foreigner, but the official system uses a succession of periods which now correspond to the reign of the emperor. The present emperor came to the throne in 1926, and that year is known as first year of Showa Period. Showa is the name by which the emperor will be known after his death. Counting from the year 1926, the year 1965 would be Showa 40.

Houses in Japan are not numbered in streets but are identified by the plot of ground on which they are built *(-banchi)*. These are numbered, as are the districts *(-chome)*. Larger units are *-cho* or *-machi*, which may be a small town in a country area or part of a large city. Another subdivision of a large city is a *-ku*. City is translated by *-shi*.

A department, comparable to an English county, is suffixed *-ken*. Tokyo is unique in being a *-to*, or metropolitan area. Osaka and Kyoto are classified as *-hu*, which are similar to *-ken*. In country districts there are also *-mura* (villages) and *-gun* (subprefectures). The latter is not a unit of local government, but purely a postal area.

Thus a Japanese correspondent might live at:
Tokyo-to. Shinagawa-ku, Gotanda-machi, 5-chome, 60-banchi.
He would put the name of the town first when writing in his own language. His surname would precede his given name. The date would be written with the year first, then the month, finally the day. He would begin his letter at the right hand side of the page and his kanji and hiragana would run from top to bottom. If he sends a present of a book, it must be opened at what we consider to be the back.

The difference between Eastern and Western usage serves to emphasise the fact that the two hemispheres were for so much of their cultural development cut off from each other. Things are now changing rapidly. It is just as easy to say *sayonara* as 'goodbye', or *kon nichi wa* for 'good day'. Try it!

4
The Rise of Feudalism

From the beginning the attempt to centralise the country's government in the hands of the emperor and the Fujiwara was a failure. Although the Japanese appear to adopt overseas ideas readily, they always refashion them into a Japanese mould. Instead of a pyramid of power with the emperor at the apex, the result of the political reforms was a growth of feudal estates (Shoen). The clan leaders emerged more powerful than ever. Their determination to avoid the new taxes based on land encouraged them to grab tax-free land wherever possible.

How The Estates Were Built Up
The loophole in the law that the lords noticed was that a distinction was drawn between public and private land. The original idea, of course, was that all land except the emperor's should be public, but the influential members of the aristocracy persuaded the emperor that their land should also be recognised as private property. As such it was automatically tax-free.

The lords took every opportunity of building up their estates after establishing their privileged position. They asked for rewards for special services to the emperor in grants of tax-free land. Waste ground, not taken into account by the tax laws, was enclosed. Land belonging to shrines, temples or monasteries could sometimes be acquired along with its special privileges.

Finally, it was soon realised that a man with a taxable estate might benefit by surrendering it to a lord whose land was tax-free. He continued to occupy the land, but would owe duties to the lord in return for his escape from his dues to the central government.

While the lords grew stronger, the central government's

financial position grew steadily weaker. The court was the centre of luxury and corruption, and while some lords were busy building themselves miniature kingdoms, others lived at court. It was only through a system of bribes and privileges that the central government operated at all.

The provincial aristocracy, especially in areas where the Ainu were still a danger, gathered around them armies of dependants. The lords were known as *daimyo* and their retainers became the warrior caste, the *samurai*. Buddhist monasteries also kept their armed bands, in theory for protection. A precarious balance of power was kept by the emperor, or more usually, the Fujiwara, calling on one group to protect them from another.

The samurai followed a code of behaviour, *bushido,* some aspects of which affect the Japanese even today. Everyone owed a duty to his overlord. The samurai's sole idea was to obey his lord, to die fighting in his service, and if necessary to avenge his lord's death. Death in battle was glorified, especially when obeying the call of duty.

The samurai's special weapon was the sword. He alone was entitled to wear two simultaneously. His weapons were cared for with fanatical devotion. The blade must be sharp and the owner's aim unerring. Famous swordsmen were said to be able to cut a man in two, head to heel, so cleanly that he would continue on his way for a few paces before dropping to the ground. Naturally the art of producing such superb weapons was also highly prized.

Ceremonial suicide *(hara-kiri* or, more politely, *seppuku)* was widely practised if the samurai considered he had failed in his duty, or if he were faced with the need to surrender. It was also accorded as a privilege instead of death at the hands of the executioner, if the samurai had been found guilty of a crime. Also every samurai was expected to live a life of austerity. In contrast to the softening life of the court, he was to be always ready for war. He despised luxury.

The Decline of the Fujiwara
In the eleventh and twelfth centuries the emperors made a bid to restore their own power at the expense of the Fujiwara. Instead of abdicating and enjoying a life of pleasure while the Fujiwara

uncle managed affairs for the child-emperor, the abdicated ruler continued to rule through his son, giving the Fujiwara relations no opportunity to establish their claims to guardianship. The abdicated emperors reinforced their religious position by becoming monks and ruling indirectly from the cloister.

The religious functions of the monasteries were pushed into the background during this period. Wealthy and ambitious, they even armed the monks, who took to the streets and fought each other, behaving more like samurai than men of God. The court had to call on the powerful Minamoto clan to restore order as monks brawled in the streets of the capital.

Dissension was rife among the different branches of the Fujiwara family. At the time when they most needed unity they were affected by the general desire to enlarge their own estates in the prevailing disorder.

The fights between the samurai of rival daimyo at last led to a full-scale civil war known as the Gempei War. The lords took sides with either the Taira or the Minamoto clans. In 1160 the Taira family succeeded in defeating the Minamoto faction and killed the leading members of the clan. Four sons were left, however, and the eldest, Yoritomo, aged twelve, was exiled to the Izu peninsula far in the west. The others entered monasteries. For twenty years the Taira dominated the court, making themselves enemies by their harsh behaviour.

Meanwhile Yoritomo grew up in exile. He made friends with his goaler Hojo Tokimasa, and married his daughter. In 1180 Yoritomo put himself at the head of the discontented nobles and with his monastery-trained brother, Yoshitsune, at his side, he advanced to meet the Taira forces once more. In 1183 the Taira fled from Kyoto, taking with them the boy emperor.

Yoshitsune was particularly successful as a general and the two brothers defeated the Taira on land. Taking to their ships, the Taira were pursued and suffered a final defeat at sea in the Battle of Dannoura in the Straits of Shimonoseki. The young emperor was drowned and the Sacred Sword belonging to the regalia was lost for ever, though the rest was recovered.

The Shogunate

Yoritomo was now supreme, but in order to ensure his position, he turned against his brother, Yoshitsune, whose fame as a warrior threatened to make him a rival for power. With Yoshitsune dead, Yoritomo had the lords swear allegiance to him. The Taira lands were incorporated in his own estates or granted to his loyal retainers.

In 1192 the new situation received official recognition when the cloistered emperor, who had survived the turmoils of the period safe in his monastery, gave the title of *Sei i tai shogun* or 'Barbarian-subduing Generalissimo' to Yoritomo. The title, abbreviated to *Shogun,* was not new, but now acquired a new significance.

In characteristic Japanese style, the old forms of government continued apparently unchanged, while discreet innovations were made. The court remained in Kyoto. A new emperor was installed. Even the Fujiwara family was allowed to retain its court positions. Yoritomo recognised the enervating effect of the luxurious life of the capital, and established his own headquarters at Kamakura. His administration is known as the *Bakufu* or 'Camp Office', a name which typifies the austere atmosphere the Shogun preferred.

Although the old imperial courts of law remained and the old system of government went apparently unchanged, Yoritomo established his own courts, military governors and tax-collectors. The authority of his officials was much stronger than those of the emperor's since the former had the backing of the army.

When Yoritomo died in 1199, his wife's relations saw their chance to step into power. That his ruthless behaviour had not gone uncriticised is shown by the contemporary rumour that the fall from his horse (which brought about his death) was due to the animal rearing as it was confronted by the ghosts of the boy emperor and the Shogun's dead brother.

The Hojo did not make themselves Shoguns, but were content with the title of Regent. The empty title of Shogun was held by the Minamotos until the direct line died out. Then the Fujiwaras and other members of the royal family were allowed to take it. The Hojo Regents governed the country, using the framework of authority set up by Yoritomo. In the background were the puppet emperors controlled by their cloistered fathers.

The Mongol Invasions

In 1274 and 1281 AD the Mongols under Kublai Khan, who had overthrown the Sung dynasty in China, pushed on to attack Japan, via Korea.

They first ordered Japan to acknowledge Mongol overlordship, but this was refused. The invasion which followed was successful in taking the main islands off the coast of Kyushu. From there the Mongols launched an attack on the north-west coast of Kyushu which was met with traditional samurai bravery. Kublai Khan found he was up against a more stubborn enemy than he had encountered so far, and when storms threatened to disrupt his supply lines, he withdrew his troops.

Since the Japanese had suffered very heavy losses, the Mongol conqueror felt that they might now be ready to agree with his terms. He sent ambassadors to suggest that allegiance and a small tribute was better than continual bloodshed. The messengers were summarily executed.

Infuriated, the Khan prepared a great fleet, which sailed for Kyushu in 1281 AD. A foothold was gained on the coast once more, but again the weather came to the rescue. A fierce wind tore the fleet to shreds. Japan had been saved again by the *Kamikaze*—the 'Divine Wind'. Encouraged by this sign that the gods were with them, the soldiers slaughtered the Mongols and forced them to withdraw, never to return.

Effects of the Mongol Invasions

The memory of divine intervention has been kept alive in Japan until the present day. The Kamikaze pilots of the Second World War were to have swept back the invading Americans as the typhoons had swept away the Mongols.

The Japanese now realised the vital geographic position of Korea for the safety of Japan. It was said to be 'an arrow pointing at the heart of Japan' and from the time of the Mongol invasions occupation of the peninsula was viewed with suspicion.

Success over the Mongols confirmed the samurai in their belief that they were the greatest warriors in the world. For centuries this belief was not seriously challenged by outsiders.

A more obvious effect of the war was the weakening of the

Hojo Regency. The expense of the campaign taxed the economic capacity of the government to the utmost, and armies had to be kept in readiness in case of another invasion. The battles being purely defensive and fought on their own soil, the soldiers could not be satisfied with plunder but had to be paid by the Regent. Still, another fifty years were to pass before the gathering forces of dissent overthrew the Hojos.

5
The Ashikaga Shogunate

In 1318 the emperor Go-Daigo inherited the throne. He was already thirty and a strong-willed man. His accession to power happened to coincide with the rule of a weak Regent, Takatoki, a man who preferred to enjoy the pleasures of his court, leaving his duties to be carried out by officials.

After eight years as emperor, Go-Daigo was expected to retire as his predecessors had done, but he refused. To make his position quite clear he nominated his son as his heir-apparent. Obviously he aimed at restoring the monarchy to power.

When in 1331 a plot to overthrow the Bakufu was discovered, Takatoki roused himself sufficiently to dispatch a force to arrest the conspirators. Go-Daigo was on the list, but he escaped and gathered his supporters around him in western Japan. As his general to lead an army against the rebels Takatoki made the fateful choice of Ashikaga Takauji. Instead of capturing Go-Daigo, Takauji saw an opportunity of overthrowing the Hojo Regency and substituting himself. He restored the rebel emperor to his court in Kyoto, then turned his forces towards Kamakura. When he saw that defeat was inevitable, Takatoki, the last of the Hojo regents, committed suicide with the members of his family and about 800 retainers who were with him.

It looked as though Go-Daigo could now rule from Kyoto without interference from either Shogun or Regent. However, Ashikaga Takauji had other plans. He first plotted to get rid of the heir to the throne, and was able to persuade Go-Daigo that the young man wished to overthrow his father. The prince was exiled and later murdered by Takauji's brother.

The treacherous general then began to cast doubts on the loyalty of the emperor's other supporters, but the latter became suspicious and ordered his troops out against the general.

Takauji was victorious however, and marched towards Kyoto. Go-Daigo fled, taking with him the imperial regalia. He established a new court at Yoshino to the south of Kyoto.

In 1336 Ashikaga Takauji declared that Go-Daigo had forfeited the throne. He proclaimed Kogon, a descendant from a junior branch of the imperial line, as rightful emperor. Two years later his nominee gratefully bestowed on him the title of Shogun. The new Shogun's court was established in the Muromachi quarter of Kyoto. Hence the era is known as the Muromachi Era and the Shogunate is known as the Ashikaga or Muromachi Shogunate.

There was continual warfare between the rival emperors, but Go-Daigo never succeeded in rallying the country behind him, although the 'Northern Emperors' reigning in Kyoto are regarded in Japanese history as the 'false emperors'.

Warfare continued after the deaths of the first contenders, but in 1392 a typical Japanese compromise was reached. The southern court gave up the imperial regalia to Emperor Go-Komatsu of the northern dynasty, while the southern dynasty emperor was adopted as his father.

The Japanese, who like to quote historical precedent, compared the Meiji restoration of 1868 to the restoration of power to the Emperor Go-Daigo. Actually the comparison is weak, since Go-Daigo's success was so fleeting.

The decision to establish the court of the Shogun at Kyoto was undoubtedly a mistake, though Takauji probably thought he would be at hand to watch for any plotting by the emperor. Kamakura had been a strategic position for controlling the important area known as the Kanto (around modern Tokyo). The Ashikaga Shoguns made some attempt to hold the area through officials, but inevitably lost their power in the countryside because of delegation of authority.

Takauji realised the need for living a simple life and preserving a vigorous soldiery. He copied Yoritomo in laying down a list of rules to be followed by his supporters, but he himself failed to set a good example. He was soon copying the way of life, self-indulgent and luxurious, displayed by the royal court.

The Muromachi era is notable for the development of Japanese art forms under the influence of renewed contact with

China. Painting (especially on religious themes), porcelain, faience, lacquer ware and damascening were encouraged. The two courts provided a continual demand for objects of beauty. *Noh* drama developed at this time too, for entertainment at the courts.

As the Ashikaga period continued the imperial court declined in wealth and importance. The Shogun became a patron of art, a very different figure from the stern warrior he should have been if he wanted to maintain his position.

By the fifteenth century the country was rent by civil wars. The daimyo were completely out of control and had collected their private armies to make war against each other, unperturbed by the feeble orders that issued from the Shogun. The only hope for a peasant or small landowner was to gain the protection of a daimyo. Private landowners were becoming fewer and fewer as they made over their land to the daimyos or were squeezed out of possession by inability to meet taxes. Many became *ronin*—landless vagrants—attacking travellers and haunting the crowded alleys of the growing towns.

The unsettled conditions encouraged the growth of towns around feudal castles, where large bands of retainers required goods and services, and where protection could be afforded to the markets. Monasteries and temples also attracted traders who provided food and other services for the worshippers. Militant monks performed the same protective duties as the lords' retainers so that settlements also grew up around the popular religious establishments.

During the civil wars the peasants occasionally got so impatient with the continual, senseless warfare of the daimyo that they banded together to defend themselves. They were goaded into action when opposing armies trampled their crops and the roads were unsafe for months on end, while the tax officers continually presented demands for more rice. Of Kii province, an old record states that it 'rid itself of the local magistrates and the farmers were the masters of the province'. The local officials whose job was to collect taxes for the lord collected more than their due and pocketed it themselves. United under the cry of 'Down with the samurai', the peasants evidently sometimes succeeded in freeing themselves, for a while at least, from the oppression.

Pawnbrokers and *sake* (rice wine) merchants were frequently the target for the peasants' anger. The former kept big storehouses filled with rice and household goods that they had accepted from the peasants in return for credit in the Chinese coinage, which was becoming more prevalent. Sake merchants also functioned as money lenders. Rice itself was still the main means of exchange and measure of value, but its perishable quality and the unpredictability of the harvests often found the peasant without rice to pay his taxes. At such times the pawnbroker had to be approached and paid somehow. Frequently a farmer had no choice but to meet his debt by giving one of his children into slavery. Small wonder that the moneylender was hated and his store houses plundered.

The Nichiren sect of Buddhists took up the cause of the poor against oppression. They called for a more spiritual approach to religion, deploring the wealth of the monasteries and the worldliness of many of the monks. Their words found a ready response amongst the peasants living on monastery land, who were just as oppressed by their abbot landlords as neighbouring peasants were by their daimyo overlords.

The Ashikaga Shoguns desperately sought any means of raising money. The rice collection from public lands was largely in the hands of the daimyo. In desperation they looked to the new, despised merchant class for supplies. Merchants were forced to make 'voluntary loans' to the Shogun, which might, or might not, be repaid. The merchants were, however, given every encouragement to develop trade with the Chinese mainland, and despite everything, they grew steadily in wealth. Japanese ships were pushing further and further afield. Lesser nobles and merchants combined in their own interests and financed trading ventures, which combined legitimate trade with piracy to the benefit of their patrons.

Although the period was one of continual unrest, the Ashikaga Shogunate survived for approximately three hundred years. The intial mistake was in letting the court establish itself in Kyoto, but the régime was doomed when it allowed economic power to pass into the hands of the daimyo. Most of the Shoguns exercised no more authority than did the emperors.

6
Religion

Like other cultural aspects of Japanese life, religion has been influenced by Chinese examples.

Originally the Japanese were animists, believing that every aspect of nature had its spirit. The idea still persists in *Shinto*. The belief that the gods are the ancestors of the whole race and especially that the Sun God was the ancestor of the emperor, has already been noted. Many of the festivals still celebrated in Japan are very ancient and are full of references to events in the legends. In country areas rain-making and fertility ceremonies survive.

The Chinese began as polytheists, worshipping many gods—the spirits of heaven and earth, departed ancestors and natural phenomena, such as trees and waterfalls.

Confucianism
About 550 BC Confucius (Kung Fu Tzu) was born. He is no legendary figure and is said to have descendants living today. His philosophy was to find wide acceptance in China and to have great influence on Japan.

Confucius stressed moral rather than religious messages. Man's duty consists in preserving a correct relationship with his fellows. Unlike the Christian thesis that man has to overcome an innate tendency to relapse into sinful behaviour, Confucius considered that man had only to obey his conscience to live a perfect life. His teaching was conservative and stressed the idea of obedience to authority, either of the ruler or, in family life, of the father. So little mention is made of God that Confucius tends to be regarded as a philosopher rather than a religious man. A more spiritual approach was supplied by the teachers of Buddhism.

Buddhism

Siddhartha Gotama, Buddha, was born in North-East India in 560 BC. The son of a nobleman, he married and had a son, but at the age of twenty-nine he left his home and set out to search for religious revelation.

There is a resemblance between the stories of Buddha's life and those of the life of Christ. This may be due to later incorporations into the spoken tradition of his adventures, which was all that existed at first. However, unlike Christ, Buddha lived to be an old man and died a natural death, and during his lifetime he did not found a church. He walked the roads in his native India, accompanied by his disciples, receiving food and lodging from anyone who would provide it. His time was spent in preaching and in meditation, and his message was for rich and poor, women as well as men.

Buddha taught that by right living everyone could eventually be freed from the necessity of rebirth into this world. The Buddhist believes that his soul must go through a series of rebirths until it achieves perfection, when it will be reunited with the spiritual essence from whence it came. God is not visualised as a person and it is impossible to convey the Oriental beliefs in the same terms as those used in the West. Both Christians and Buddhists believe that their life and how they cope with its problems will affect their future. The difference is that the Buddhist is sure that he will come back to this earth to receive his reward or his punishment. Only someone who has achieved perfection can hope to be freed from the weary cycle. Buddha advocated the 'Middle Path'—moderation in all things. He did not even claim to have the only answer to salvation, but felt that everything, including evil and misfortune generally, has a place in the scheme of things. The Christian problem of how God can tolerate evil does not trouble the Buddhist.

From such simple beginnings grew up a mass of ritual, temples, monasteries and images with many different interpretations of Buddha's teachings. Monasticism has become very widespread. In Tibet monks used to make up a large proportion of the population. The religion there had incorporated pagan deities and the believers often claimed magic powers, fostered by their complete mastery over the body. Heads of monasteries and the Dalai Lama, the priest-ruler of Tibet, were claimed to be the reincarnation of the previous occupier of the position. As one died his soul passed almost immediately into a newborn baby, which could be identified by a series of special signs.

In Thailand it is the custom for all adult males to spend at least a few months as a Buddhist monk.

In all countries where Buddhism has been adopted differences show themselves and aspects of the earlier religions of the district have been incorporated. In China and Japan the influences of Confucianism and Shinto can be traced.

Buddhism was introduced to China in the sixth century AD and thence, via Korea, to Japan. Here it existed side by side with polytheistic Shinto.

Later Religious Developments in Japan

Amida: This sect first appeared in India about the first century AD and spread thence to China and Japan. Amida is a Bodhisattva: one who by good living has qualified to cease being reborn, but who elects to remain on earth to help others. He is therefore a sort of saint. His sect is distinguished by the belief that repetition of Amida's name will bring salvation.

Tendai and Shingon: Sects of the ninth century AD, these both make a feature of ceremony. They stress that everything in the universe partakes of the Buddha nature. In Shingon particularly this leads to images depicting the more unpleasant side of life, instead of the benevolent saints usual in Japanese Buddhist art.

Nichiren: at the end of the thirteenth century Nichiren set himself up as a sort of protestant reformer of Buddhism. He called for a return to the teachings of the master as he felt they were expressed in the sacred books.

Zen: Zen Buddhism has attracted much interest in the West and is very widespread in Japan. It teaches that enlightenment will come suddenly and in a flash as the devotee follows the Zen technique. Inaction is seen as quite as useful as action. From this idea comes the practical application of Judo, where the fundamental principle is to let one's opponent overcome himself by his own force.

Zen Buddhist monks live by begging, and all share the work

of the monastery. Meals are very simple and are eaten in silence; most of the day is spent in a hall where each monk is allotted a space three feet by six. He sits motionless in meditation. Periodically he reports to his teacher, who sets him topics for meditation and tests his progress. The devotee is asked to solve riddles which appear to be without an answer. During meditation he may be given a sharp blow, since it is felt that a shock of this kind may be the sudden stimulus that will enable him to overcome human limitations and achieve that insight into the supernatural that is the aim of his exercises.

The monk spends the night in the same place in the hall where he has sat during the day, and may cover himself only with a wadded cotton quilt. He is free to leave the monastery whenever he wishes to seek another teacher or to return to the everyday world.

Japanese arts like *ikebana* (flower arranging) and *bonsai* (cultivation of miniature trees) which have become popular in the West, have their origin in religion. The oriental idea of the universal nature of God, pervading all creation, leads to a desire for symbolism of this unity. A Japanese flower arrangement symbolises the heavens, man and earth. It is arranged on three levels for that reason. Bonsai enables a miniature landscape to be achieved in a very small space, and space is at a premium in the crowded cities of Japan.

The famous tea ceremony has a Zen background, being introduced originally from China. Tea is dispensed with elaborate ceremony. The whole effect should be one of tranquillity after ritual purification. The communal meal is a frequent religious symbol, and the tea ceremony should be seen as such, rather than, as some visitors describe it, a fanciful and tedious way of serving an ordinary cup of tea.

7
The End of the Civil War

Japan was retrieved from the chaos of civil war by the emergence of three great leaders.

Oda Nobunaga (1534-1582) was only fifteen when his father died and left him a small estate in central Japan. Naturally the neighbouring landowners thought this would be a grand opportunity to add to their own land, but even at that age Nobunaga proved himself a capable general. He not only held them off, but made himself master of the province. Gradually he restored order in the central area. In 1567 he entered Kyoto, but he preferred the reality of power to empty titles and did not disturb the emperor or the shogun. They continued as mere figureheads.

Order was restored in the greater part of Honshu as it came under Nobunaga's control. Some of his fiercest opponents were the monastic armies, especially among the Amidaist sects.

It is thought that Nobunaga's favourable reception to the first Portuguese traders, who came to Japan in 1543, was partly due to their advocacy of Christianity, which he saw as a new religion to undermine the power of the Buddhists. He was interested in the ships and firearms of the newcomers, too. A new fleet, built on Portuguese advice, and new weapons bought from them, played no little part in his successful campaigns.

The powerful western clans of Choshu and Satsuma were still holding out and Nobunaga was engaged in subduing them when he was murdered by one of his generals, Mitsuhide. This ambitious man seized Kyoto and forced the emperor to nominate him as shogun. He killed Nobunaga's heir, but the two closest associates of the dead general were still at hand. These were Toyotomi Hideyoshi and Tokugawa Iyeyasu.

Hideyoshi was the son of a poor woodcutter, who saw his chance to rise by attaching himself to Nobunaga and had become his most trusted lieutenant. When Nobunaga was killed Hideyoshi was fighting in Choshu. He returned in a series of forced marches and slew all concerned with the murder. He named Nobunaga's infant grandson as his successor. Two other sons of Nobunaga, passed over because their mothers were only concubines, tried to assert their claim to their father's possessions and power. For a while they were supported by Iyeyasu, their late father's other close associate; but he soon realised that his best interests would be served by returning to co-operation with Hideyoshi.

The Work of Hideyoshi (1582-1598)
Hideyoshi continued Nobunaga's aim of bringing all Japan under the suzerainty of one ruler. He overcame all the outstanding feudal lords in Shikoku, Kyushu, the Kanto and northern Japan. Even Satsuma and Choshu were forced to give allegiance.

He continued the attack on Buddhist power and at first followed his predecessor's policy of friendship to the Portuguese, but when the Buddhists had been suppressed, he had no desire to see the Christians take their place. In 1597 he ordered the execution of Christians and the expulsion of missionaries.

Hideyoshi established Yedo, now Tokyo, as his new capital, leaving the emperor and the shogun to rule, in their fashion, from Kyoto. While they occupied themselves with pomp and ceremony, he was the real power. He was a great castle builder and many can be seen today, some restored to their original massive splendour. Osaka castle is the best known example. When all the feudal lords had been defeated in 1590, Hideyoshi would have liked to have become shogun, but his humble birth was held against him. He finally persuaded one of the branches of the Fujiwara family to adopt him and was then deemed eligible for the title of Kampaku, or Dictator.

With law and order restored, the country could settle down to develop trade and industry. Hideyoshi viewed with alarm the tendency, fostered by the disturbed times, for peasants to leave their land and for people in general to forget their position in feudal society. His own career is a good example of the breakdown in the rigidity of the system. Laws were issued forbidding peasants to continue their drift into the towns. Weapons were confiscated from all except samurai. Everyone was expected literally to mind his own business. Taxation and coinage were reformed and an attempt made to bring about an equitable redistribution of land.

In 1592 Hideyoshi attacked Korea. His aim was to use it as a stepping stone towards the conquest of China. The armies were successful, but three factors brought about the failure of the expedition. The Japanese supply line was broken by a brilliant Korean admiral, who used iron-plated ships to ram the wooden ships of his enemy with devastating effect. Secondly, the Chinese poured troops in from the north to help the Koreans. Finally, the bitter winter weather and the mountainous terrain proved too much for the invaders.

A second attack was made in 1597 and was meeting with a similar lack of success when Hideyoshi died in 1598.

He had hoped that his young son, Hideyori, would succeed him, and before his death he tried to persuade his former comrade in arms, Iyeyasu, to promise to rule on behalf of the boy until he came of age. However, Iyeyasu was reluctant to take oaths which would tie his hands in the future. Hideyoshi had to be content with appointing a committee of Regency to guide his son. Inevitably the members soon were at loggerheads with each other and Iyeyasu had to step in. Those who took up arms against him were soundly defeated at the Battle of Sekigahara in 1600.

The estates of Iyeyasu's enemies were confiscated and reallotted to his friends, whose territories were craftily mingled with those of the less reliable lords' lands to guard against a hostile alliance being made. Iyeyasu could not forget that Hideyoshi's son still existed as counter-claimant to his authority. He consolidated his position by marriage alliances with Hideyoshi's family, who themselves were allied by marriage to members of the leading families.

In 1603 Iyeyasu was given the title of shogun, and founded the Tokugawa Shogunate, which was to last till 1868. He remained on good terms with the Emperor, but stressed his religious aspect and belittled his political functions.

To put a further curb on the daimyo, Iyeyasu ordered that each of them should maintain a residence in Yedo, now the shogun's capital. Some member of the family must always be in residence as hostage for the good behaviour of the lord. The head of the household must pay homage twice a year to the shogun. The effect on the city of this legislation was a phenomenal growth of buildings to house the lords, their families and retainers. Shops and places of entertainment sprang up to meet the needs of the new population.

The roads to Yedo were always thronged with travellers making their compulsory visits to the capital, and along the routes were gradually established inns and towns to cater for the overnight accommodation of the nobility and all the servants they thought it necessary to display their importance.

In 1605 Iyeyasu retired from the shogunate in favour of his son Hidetada. Like the cloistered emperors, he maintained his authority, but he retired from Yedo and set himself up in Shizuoka, 120 miles away.

Iyeyasu and Hidetada were becoming increasingly suspicious of the motives of the Catholic missionaries who were making converts among the Japanese. Dutch and English traders and men like Will Adams (see Appendix 2) brought with them memories of the religious feuds of Europe. They hated the Spaniards and Portuguese both for their religion and as rivals in trade.

In 1614 Christianity was once more declared outlawed and this was the signal for many Japanese Christians to join Hideyori at Osaka. Masterless samurai, cast adrift after the battle of Sekigahara, had for years been urging the young man to claim his birthright. Iyeyasu's first attack on Hideyoshi's son, defending himself in the castle at Osaka, was a failure. A second attack in the following year was successful after a siege of two months. Hideyori's young son escaped, but was later captured and killed. Hideyori and his mother preferred suicide to capture. Their dead bodies were found among the corpses of the warriors who defended the castle to the end.

In 1616 Iyeyasu died. For a brief period it had looked as though Japan would take her place with the rest of the world, becoming involved as did the other countries of the area with the European adventurers who were eager to extend trade in the sixteenth and seventeenth centuries. Iyeyasu's descendants thought differently.

8
European Interest in the East

Despite the distances involved and the difficulties of travel, contact between East and West has been maintained since very early times. Goods from the East found their way by overland routes to the shores of the Mediterranean. Greek ships are known to have called at Indian ports.

For a long time trade connections were in the hands of the Arabs. From their strategic position across the trade routes the Arabs were able to enjoy a monopoly of such Eastern products as camphor, cloves, sandalwood and nutmeg.

Too expensive for the average man, the spices and silks from China and the Spice Islands were eagerly sought after by the wealthy. Silk was imported from China by a perilous overland route through the famous cities of Bokhara and Samarkand. Then in about 550 AD some of the silk worms were smuggled into Europe and were carefully tended until their offspring were able to produce enough thread to satisfy the European market. The fabric was still literally 'worth its weight in gold', but the supply was no longer dependent on the survival of caravans from the East.

In 1275 the Italian traders, the Polos, made their way to China. The famous Marco Polo spent seventeen years there before returning home by sea. He was entrusted with a Chinese princess, who was to become the bride of a Persian prince. She died before the long journey was completed.

Christian missionaries regularly visited China until the Ming dynasty cut off contact with the West in the fourteenth century. Their teachings seem to have made little impression, but the Arabs were more successful. Marco Polo, held up for several months on his way home from China, had time to note the prevalence of Mohammedanism in Sumatra. The seeds were already sown for the strong Moslem groups in present day Indonesia and Malaysia.

The Moslem drive eastward was matched by a thrust westward through north Africa to Europe via Spain and Portugal. From their Moorish neighbours the inhabitants of the Iberian Peninsula learned of the wealth to be gained from voyages to the East.

Early in the sixteenth century the Portuguese discovered the route around the Cape of Good Hope. In 1510 they captured Goa in India, and the following year they took Malacca. The Portuguese aim was monopoly of the spice trade, so they were far from pleased when a Spanish expedition under Magellan arrived at the Spice Islands, as the Indonesian islands were then known. In 1493 the Pope had divided the new world into two spheres of interest between Spain and Portugal. He had not foreseen that Spain might continue west of the Americas, which were supposed to be her territory. The Portuguese claimed that any islands which might be discovered in the Pacific were theirs by virtue of the Pope's decree, but the Spanish were eventually successful in substantiating their claim to the Philippines, which were named after the Spanish king's son.

The Arab influence was meagre in the Philippines and the Spaniards made a determined effort to convert the people to Christianity, with the result that the islands are now predominantly Roman Catholic.

Spain and Portugal were united under one ruler in 1580. In theory they were no longer rivals, but in practice brawls were still frequent among traders.

The English were introduced to the possibilities of lucrative trade with the Pacific area by their enmity for the Spaniards. Sir Francis Drake continued westwards into the Pacific after attacking Spain's South American possessions. On arrival at Ternate (Indonesia) Drake was welcomed by the Sultan, who thought he might use the newcomer against the Portuguese. The cargo of cloves that the English brought back with them in 1580 convinced the English merchants that here was a field into which they might most profitably venture. As relations between Spain and England deteriorated, damage to Spanish trade would

The Beginning of European Interest in the East

be patriotic as well as profitable. In 1586 an expedition under Thomas Cavendish reached Java, and in the following years English adventurers combined trade with attacks on Portuguese and Spanish shipping and trading posts. In 1600 the British East India Company was formed. Its aim was to make as much profit as possible by trade in the East. Spreading Christianity was none of its concern.

The fourth country to become involved in the Pacific area was Holland. Although that country was not officially recognised as freed from Spanish sovereignty till 1648, its actual independence had been established at the turn of the century. The Dutch East India Company was formed in 1602. Although they shared a common enmity for Spain, the English and Dutch were now trade rivals. Matters came to a head in 1623 in what is known as the Massacre of Amboina. The members of the English trading post there, among whom were eleven Japanese, were accused of plotting to seize the Dutch fortress. They were tortured and 'confessed'. Ten English and ten Japanese were executed. As a result the English withdrew and left the Dutch to establish their control over what became known as the Dutch East Indies, now Indonesia. English trade centred on what is now Malaysia and on India. The Portuguese were pushed out until only odd islands, Goa and Macao, remained, and Spain concentrated on the Philippines.

European Contact With Japan

In 1543 Portuguese traders from Macao were driven off their course and landed in southern Japan. They were welcomed by the local daimyo, who bought guns from them. Succeeding expeditions were well-received especially with the new weapons which could be used in the warfare between Nobunaga and his enemies.

In 1549 St Francis Xavier and two Spanish Jesuit priests landed at Kagoshima, the stronghold of the powerful Satsuma clan. He then moved to Hirado where the Portuguese had established a trading post. Here the respect shown him by the Portuguese impressed the natives, some of whom were converted. However, after a trip to west Honshu and a walk from there of 300 miles to Kyoto, Xavier realised that his humble garb and

Christian simplicity were interpreted by the Japanese as lack of consequence.

He returned to Yamaguchi in style and bearing expensive gifts from Portuguese dignitaries. This time he was received by the local Japanese nobility and was able to deliver his Christian message. Unfortunately he never fully mastered the language and his violent attacks on corruption and moral laxity were not welcome. His successors were more tactful, and the Japanese were glad to learn Western ideas from them.

Under the protection of Nobunaga, priests were even received at court and Christianity made some headway in Kyoto. At Nobunaga's death their number in Kyoto was estimated at 10,000. In western Japan the estimate was 150,000.

Hideyoshi viewed the new religion with distaste. The Portuguese were trying to gain complete control of Nagasaki, and were threatening to withdraw their trade if they were not granted their wish.

The new converts, as is often the case, were more keen than their instructors and showed an intolerance of the established religions that was strange for Japan, where the sects might attack the government but seldom each others' views.

Hideyoshi also disliked the influence of the Catholic priests, who urged their converts that their first allegiance should be to their religion and to the Pope. Obedience to an outside power could prove much more dangerous than obedience to a rebellious daimyo.

Jesuit missionaries were ordered to leave Japan in 1587, but the edict was not rigidly enforced.

During the last eight years of his life Hideyoshi seems to have been obsessed with a desire to extend his power. In 1591 he sent messengers to the Philippines claiming suzerainty, relying on the report of a returned Japanese adventurer that the Spaniards were weak. The Spaniards instead sent Franciscans to offer trade and ask for tolerance for missionaries. They were inclined to be conciliatory because the Japanese had already shown belligerence by their attack on Korea, and the Spaniards knew that apart from their sea power they were not in good shape to defend the Philippines. Whether deterred by his lack of success in Korea or by the threat of Spanish sea power, Hidey-

The Beginning of European Interest in the East 43

oshi did not follow up his claims, but listened favourably to the Spanish suggestions.

In 1597 he turned against the Christians again, executing many and driving out the priests. The last straw appears to have been that Spanish sailors were heard boasting that after the traders and missionaries would come the troops of Spain to annex Japan.

The new Shogun (1600), Iyeyasu, was more tolerant at first. He saw the trading benefits the foreigners were bringing to his country, and tried to get them to use the port of Uraga, near his capital, instead of concentrating on Nagasaki.

Up to the death of Iyeyasu the Japanese showed every sign of taking their place as the great sea faring and trading nation of the east. Japanese mercenaries and adventurers were everywhere and often influential. Shipping increased under the guidance of Iyeyasu and his English advisor, Will Adams.

The expansionist policy also showed itself in the desire to push the Japanese frontiers past the home islands. In 1608 the king of the Ryukyu Islands was ordered to pay tribute, and an attack in the following year forced him to concede.

In 1632 the Japanese attempted to acquire a foothold on Taiwan (Formosa), where the Dutch had established a fort, but they were repulsed.

Many of the daimyo were actively encouraging trade. The daimyo of Sendai, in northern Japan, sent an embassy to Rome in 1612, offering to give a favourable hearing to missionaries in return for trade. This was in defiance of the anti-Christian edicts put forth that very year by Iyeyasu.

The Shogun saw that Christianity was a real threat to the peace of his dominions. The Europeans did not scruple to play off one daimyo against another and use their cupidity for their own ends. The Counter-Reformation in Europe had brought about a resurgence of aggressive Catholic proselytising, which found its echo in Japan. Protestant merchants voicing their hatred and fear of the Jesuits found many sympathetic listeners among the Japanese.

In 1614 all foreign priests were ordered out of Japan, churches were demolished and converts were instructed to renounce their faith. Iyeyasu's son began a ruthless persecution of Christians, but in 1636 his grandson, Iyemitsu, took the final step of forbidding the entry of any foreigners into Japan. Only the Dutch and Chinese were allowed to land on the island of Deshima in Nagasaki harbour for the purpose of trade. The Dutch were permitted to send one ship a year from Holland, and once a year the Dutch merchants were permitted to emerge from Deshima to go to Tokyo to present gifts to the Shogun. The court nobles were encouraged to jeer at the Dutchmen and humiliate them by forcing them to perform undignified acts, but all was deemed worth the lucrative trade that was the Dutch monopoly.

No Japanese was henceforth to leave the country. Those who were already abroad were forbidden to return on pain of death. Even ships and sailors were covered by this edict.

In 1638 a final edict was issued banning the foreign religion. The Christians around Nagasaki rebelled and took possession of the fortress of Shimabara. Doubtless they expected help from their co-religionists, the Dutch, who, however, thought it more in their interests to assist the Shogun by bombarding the helpless defenders from the sea. The fort was taken and thirty thousand Christians were massacred.

Thus began the seclusion period of Japanese history. Trade with China was maintained, but only a thin trickle of information from the West penetrated Japan until 1853, when Commodore Perry arrived from America. The ban on Christianity was to remain until 1873. When it was officially removed it was revealed that around Nagasaki the religion had been secretly kept alive and practised for two hundred years.

9
The Seclusion Period

For approximately two hundred years Japan was virtually cut off from the rest of the world and for this period she was involved in no external warfare, so that her history is solely social and economic. Culture was frozen. Perry visiting the islands in 1853 was like a time-traveller stepping back over the centuries.

To quote a Japanese economist, E. Honjo: 'The rise of the Tokugawa Shogunate is attributable to its possession of financial power far superior to that of many feudal lords, while its downfall was due to the fact that the financial basis of the Shogunate was shaken as land economy was supplanted by currency economy and also that it proved incapable of controlling the economic power of the rising shonin [merchant] class.' This statement can be criticised as an oversimplification, but it is a neat summary of the economic reasons for the rise and fall of the Shogunate.

Until the Meiji restoration there was no attempt to separate the public from the private finances of the emperor or shoguns. The former managed to live well off their extensive estates in the main, though they were subject to the vagaries of the seasons. Japan had a rice economy. Dues to the nobles and emperors alike were paid mainly in rice and a bad season was liable to affect even the Court. The Taika Reform of 646 was, as we have seen, an attempt to broaden the basis of the government's economy, to give it a taxation system administered by a bureaucracy. If it had been successful the government of Japan must have developed along different lines.

Instead the landowners evaded the reorganisation and by establishing the *Shoen* they evaded the taxation, and laid the foundation for the feudalism typical of Japan. It should be noted that although the same term is used, the organisation of feudalism in Norman England resulted, for a while at least, in a pyramidal structure with the king at the head. In Japan, on the other hand, the system was from the first founded on the determination of the landowners to maintain their economic independence. How far they were controlled by the central government was always dictated by the military power it was able to command, a power which in its turn rested on the economic strength of the ruling family.

As the shoen grew, the burden of taxation on the peasant working the 'public lands' (the only taxed land), became heavier. Many were forced to leave and seek a living in the towns which were growing up. Markets were established under the protection of leading temples where the throng of worshippers provided customers. These formed the nuclei of towns, as did the feudal castles. Trade demanded stronger security, which was provided by the lord, whose retainers were ready customers for both commodities and entertainment.

The use of coins as a means of exchange was introduced as early as the Nara period, but was mainly limited to small amounts imported from China and supplied as a balance of trade between the two countries. Some coins were minted in Nara and Kyoto and used locally, but there was a tendency to hoard coinage because of its durable quality, in marked contrast to the usual means of exchange, rice.

Nevertheless, this perishable commodity was collected in large quantities by the lords, the emperor and the shogun and had to be housed in storehouses controlled by officials. Before it deteriorated it had either to be used or exchanged for more permanent goods. Inevitably the officials in charge of the big landowners' storehouses became influential and wealthy themselves. The famous merchant house of Mitsui owes its early start to such an occupation.

Sake (rice wine) merchants and pawnbrokers were others who, because they had access to more durable goods, were able to rise to wealth by offering loans at a high rate of interest, to be paid usually in rice. These people were in a position to take advantage of fluctuating supplies. If natural disaster struck some farming area, a very frequent event, the local rice capitalist was ready to take his profit.

The peasants inevitably fell more and more into debt and the drift into the towns became such a pressing problem that frequent orders were made by the government to the effect that farmers who left their land and neglected their farmwork were to be punished. Even though Japan had no foreign wars during the Tokugawa period the population actually declined. Apart from the effects of famines and epidemics, the people themselves practised infanticide and abortion, because their poverty would not allow them to bring up a reasonably-sized family. The daimyo naturally viewed their dwindling labour forces with alarm and some of them even introduced a kind of child endowment system, giving concessions to peasants who maintained families.

The government during the seclusion period tried to enforce a rigid social system, which however contained the seeds of its own decay. Society was officially divided into classes: samurai, farmer, artisan, merchant, in descending order. Within the samurai class were further divisions, which were laid down by Iyeyasu after the battle of Sekigahara in 1600. Those who had opposed him had their estates confiscated or reduced and his supporters were rewarded from the proceeds. At the top of the social scale were Iyeyasu's three sons, who were given estates around Tokyo and Kyoto. They founded families which were known as the *Go-Sanke,* the 'three honourable houses'.

Below them were the *Fudai,* the nobles who had supported Iyeyasu at Sekigahara, and the *Tozama,* the 'outside nobles', who had hurried to swear fealty after he had demonstrated his success in that battle.

Below these were the daimyo and lesser landowners and landless warriors, who were entitled to samurai rank from circumstances of birth. Everything was strictly regulated according to social position, dress, order in processions, the gateway to be used when visiting the court, the nature of the offerings to be made to the shogun, and so on.

As the farmer found it impossible to provide for a family on a limited plot of land, so a samurai found himself struggling to support his children and his retainers. Both classes were denied any progress by the rigid economy. Only the new merchant class, at the very bottom of the scale, was able to expand and prosper. No profession or trade was supposed to be suitable for the samurai except that of warfare, and during the Tokugawa peace he was therefore condemned to inactivity. Like the farmer, the samurai was soon borrowing from the despised merchant class.

The Tokugawa were not blind to the growing gap between poor and wealthy. Peasant revolts were sometimes frightening reminders of the unrest of the Muromachi era (1338-1573) when rioters looted the storehouses of the sake-merchants and the pawnbrokers. By edicts the shoguns tried to keep the price of rice reasonably stable, but such measures were impossible to enforce.

Equally useless were the attempts to enforce a moratorium on debts. All debts were to be forgiven and everyone was to start afresh. This had been a practice as far back as the thirteenth century and even then had caused chaos by frightening the money-lenders into withdrawing their funds. Under the threat of a *Tokusei* (debt repudiation) edict, the merchants became reluctant to lend or else they enforced agreements which would evade the edict.

The shoguns themselves sometimes issued loans to impoverished samurai. Iyeyasu had advised his successors to husband the sources he had left them and keep a store of gold 'for possible military campaigns, for relieving people in times of calamity and to provide against crop failures'. It is obvious then that the Shogun had no idea of the change to take place in the economy. He left a country with expanding trade and every sign of vigorous growth. His successors struggled on with no new source of revenue and ever growing expenses, until they were faced with the final blow of the need for defence against foreign attack.

Only the shonin, the new merchants, were prospering, and despite all edicts to the contrary, the farmers and samurai were creeping into their ranks, marrying their daughters or being adopted as their sons. Even the daimyo were beginning to see the benefits of trade. Those whose territory was favourably situated for contact with the mainland were able to sell the products of cottage industries from their own estates and monopolise the redistribution of imports.

The feudalistic system had extended to the artisan classes in the form of guilds. They too had a rigid hierarchy, of apprentices, journeymen and mastercraftsmen. They attempted to regulate prices in their own favour and to control working conditions. They were held to blame for rising prices and were very unpopular with the mass of the people. At the beginning of the nineteenth century the guilds were abolished, but the result was only to add more confusion in the economic structure of the country.

Clearly nothing short of a complete reorganisation of the system could save the tottering financial structure. Meanwhile all the Shogunate could think of was to tap the merchants' wealth by means of so-called 'voluntary loans'. In theory the money was to be paid back, but in fact it seldom was.

The government also gained some temporary relief by minting its own coinage, using the gold and silver from the mines under its control. Unfortunately, the favourite way to increase this apparent wealth was by debasing the coinage, making a new issue containing a smaller quantity of precious metal than the previous one. The daimyo also issued private notes and coins, which added to inflation.

Iyeyasu's edict that feudal lords must visit the capital regularly caused them to be perpetually on the move. The main roads were thronged with travellers who needed something easier to carry than a bag of rice to pay for their lodgings.

Although Japan was cut off from overseas contacts in the main, there was still trade with China, itself isolated from the West. Some Dutch influence filtered in through Nagasaki. Yoshimune (1716), the eighth Shogun, lifted the embargo on Dutch books. A few scholars continued to learn Dutch, study the books and make the results of their labours available in Japanese. Only books on religious matters were still forbidden completely. Developments in chemistry, medicine, astronomy and military science were followed by the devoted few.

Towards the end of the eighteenth century increasing numbers of unwelcome Western ships tried to gain entrance to Japanese ports. The government also heard rumours of Russian interest in the islands to the north, but an expedition to Sakhalin found no trace of them. The Japanese commander returned after setting up a notice claiming Japanese ownership of the land.

In 1808 a British captain foreshadowed the high-handed actions of Commodore Perry by sailing into Nagasaki, ostensibly chasing Dutch ships. He threatened to blow all the Japanese shipping out of the water, if he were not given the supplies he needed.

This and other incidents added to the Japanese alarm when they heard of China's fate in the Opium War (1839-42). In a comparatively short time China was defeated and compelled to open several ports to foreign trade.

In 1846 an American naval officer, Commodore Biddle, arrived in Japan asking for trade rights, but he was under orders not to use force and had to return home empty-handed. Not so Commodore Perry in 1853. He delivered to the Shogun a letter from President Fillmore requesting the opening of trade relations and warned that he would be back next year expecting a favourable answer.

The government of Japan began feverishly to prepare for war, building forts and buying arms from the Dutch. Leading daimyo and scholars were asked their opinion of Fillmore's letter. Some advocated a firm policy of denial of American demands, but others pointed out that Japan was by no means fit to oppose the type of weapons that European and American ships carried.

When Perry returned in 1854 he was able to procure Japanese consent to the Treaty of Kanagawa. American ships were allowed to call at the ports of Shimoda and Hakodate for supplies, and arrangements were to be made for an exchange of consuls. A most-favoured-nation clause was significant progress for the Americans, because it provided that no other nation should be granted better treaty rights than themselves.

10
American Interest in Japan

The reason officially given as America's motive for forcing Japan to open the country to negotiations was that American sailors were being badly treated when they happened to be shipwrecked on her inhospitable shores. America was developing rapidly and her ships were fast becoming the greatest users of the Pacific routes. The advent of steam power, the opening up of China and the discovery of gold on the western side of America, were all contributing factors. Sailors engaging in whaling, as well as those of ordinary merchant shipping, complained of their treatment by the Japanese. If they called into a Japanese port for water or food, they would be driven away with threats. If they happened to be shipwrecked, they were bundled off to Nagasaki to await the arrival of the annual Dutch ship.

Wider considerations affected American businessmen in their attitude to Japan. The islands were in a very convenient position to serve as a refuelling and restocking station for American ships trading with the eastern ports of Asia. Secondly, the Japanese market was as yet untapped. Enormous profits must be awaiting the first merchants to establish themselves there.

Other powers were already showing interest too. English ships had called in Japanese ports, successfully defying the bluster and medieval weapons of the inhabitants. It could only be a matter of time before they returned with a cargo of manufactured goods and a mandate to open a new trading post.

Russia's attitude seemed even more sinister. Pushing down from the north, she seemed bent on conquest. The islands to the north of Honshu, although claimed by the Japanese, were not occupied by them. Even Hokkaido was more an Ainu outpost than a part of Japan proper.

In 1849 a Russian expedition proved Sakhalin to be an island, not part of the mainland as it had been thought to be, and rumours of the activities of Russian ships reached even the ears of the Shogun. Count Muraviev, the new Governor-General of Siberia, had instructions to explore the Amur valley, review the possibility of establishing ports on the Pacific coast and generally develop the area. The Russians had established a trading company in Alaska and were competing with the Americans in whaling. In 1853 they annexed Sakhalin.

Events were moving towards the outbreak of the Crimean War and the Russians decided to try to gain Japan's friendship to circumvent the obvious interest of Britain and America. As crisis threatened in the West it would be good policy to regularise the position in the East. Admiral Putiatin's expedition to negotiate on behalf of the Russians arrived off Nagasaki only to find that Commodore Perry had got there before him. He was forced to leave Japan without accomplishing anything.

When Perry returned the following year, he brought with him a small steam train and an eighteen inch track as a present and as a further reminder to the Japanese of their technological backwardness. A miniature telegraph line from Yokohama to a neighbouring village was another cause for wonderment.

The Treaty of Kanagawa between the Americans and the Japanese was followed by a Russo-Japanese Treaty in 1855, settling the northern boundaries at a line running through the Kuriles. Russia was granted the same rights as other powers and in addition gained extra-territorial rights, that is jurisdiction over Russian nationals in Japan. Other nations later claimed and were able to obtain this concession, which became a long-standing source of irritation to the Japanese.

In 1856 Townsend Harris arrived in Shimoda as the first American consul. He was accompanied by a Dutch interpreter and the two men were given a grudging welcome by the Japanese.

It was not until the following year that Harris was at last allowed to present his credentials to the Shogun. He and his companion led a thoroughly miserable existence, since the Japanese hoped that he would get tired of waiting and go home without the full treaty he was supposed to be there to negotiate.

However, he persevered, and in 1858 a Treaty of Commerce and Navigation was accepted by both sides:
1. Ministers and consuls were to be exchanged.
2. Hakodate, Kanagawa and Nagasaki were to be opened to trade immediately. Other ports, Niigata, Kobe, Tokyo and Osaka were to be opened to trade and residence by Americans on specific dates during the next five years.
3. Duties on exports and imports were agreed on.
4. The United States was granted extra-territorial rights over American nationals.
5. Americans were to have freedom of worship.
6. The treaty should be revised after 1872.

Similar treaties were obtained in the same year by Holland, Britain, Russia and France. The Japanese later realised that they had been persuaded into granting concessions that were not normal practice. 'Tariff autonomy and abolition of extra-territoriality' became the battle-cry of the Japanese patriots.

The foreigners were still under the impression that the Shogun was the sole ruler of Japan. Actually he was by this time largely a figurehead, and powerless in the hands of groups of noblemen, who themselves were split on the best policy to follow. Some were bitterly opposed to any concessions to foreigners.

The treaties, which the foreigners regarded as triumphs of diplomacy, had actually been signed only on behalf of the Shogun with reservations known only to the Japanese. Since it suited them to do so, they could claim that the Emperor must sign also, if the Treaty was to be valid.

The Emperor Komei, who ascended the throne in 1847, was against making any concessions to the 'barbarians', as the Japanese regarded the foreign visitors. He refused to accept the treaties and informed the Shogun that the newcomers must be expelled as soon as possible.

Such was the turmoil into which the country had been thrown by overseas intervention. Behind the pomp and ceremony, which was all that the foreign diplomats and merchants were allowed to see of Japanese court life, was preparing the revolution known to history as 'the Meiji Restoration'.

11

The Fall of the Tokugawa Shogunate

The intervention of Europeans and Americans into Japanese affairs was the final factor that brought about a change of government in the country, but rumbles of discontent had been heard for some years. At the crucial moment there was no man of the calibre of the great shoguns of the past to carry the office and to push through reforms that were obviously necessary. Instead the Emperor proved to be the outstanding figure who provided the focus for the country's loyalty, and held it together in a period of great mental stress.

Reformation was needed in many fields, especially in that of economic affairs. A previous chapter has indicated the growing turmoil caused by the government's ever-increasing expenses and the inadequate nature of the taxation system.

The great shoguns had always made a feature of their policy a strict control over the daimyo. Iyeyasu had hedged them about with rules and prohibitions that had kept them quiet for the past two hundred years, but there were now signs that the system was cracking. Under Iyeyasu's orders the daimyo had to make regular visits to Yedo (Tokyo) and part of his family had to reside there always. All travellers entering or leaving Yedo were scrutinised carefully and reports were made to the shogun. Women leaving or guns going into the town were taken as an indication that a lord was contemplating rebellion.

Marriages between daimyo families could only be arranged with the consent of the shogun so that he could prevent family alliances from building up large estates in strategic areas. Iyeyasu's aim after Sekigahara had been to separate the barons of doubtful loyalty from each other by blocks of land belonging to nobles he could trust. Even these were restricted from building new castles or strengthening the fortifications of the old ones.

The shogun made sure the daimyo had no time for rebellion. They were burdened with court duties and ceremonial that took up both time and wealth. In case this was not enough, excessive accumulation of riches was curbed by the custom of calling on daimyo to repair or build temples, or the shogun's castles. What was in reality a heavy financial burden was represented as an honour and could not be declined.

No daimyo were allowed to make a direct approach to the emperor. They could not even pass through Kyoto on the way to pay their respects to the shogun. Their actions were all reported to him by an army of spies whose sole duty was to watch for signs of disaffection.

By the nineteenth century the rigidity of the social structure and Iyeyasu's precautionary measures were seriously undermined. The rule that a daimyo must spend a large portion of his time in Yedo was relaxed in 1862. His visit was compulsory only every three years and his family was no longer held hostage for his good behaviour. Undoubtedly this was only a belated recognition of the fact that the old stringent rule was not being observed and was incapable of enforcement. So was the rule that the daimyo were not to approach the emperor.

A revival of interest in the position of the emperor had been fostered by a growing body of historical scholars. Iyeyasu's grandson, Mitsukuni, Lord of Mito, one of the Go-Sanke, inaugurated an interest in Japanese, instead of Chinese, antiquities. The Kojiki and Nihongi were studied and the emperor's descent from the Sun God was given new prominence. His eminent birth and position as described in the early histories was a painful and obvious contrast to his position under the shogunate. National feeling, coupled as always in Japan, with an increasing emphasis on the national religion, Shinto, produced a climate favourable to the restoration of the emperor to prominence as the shogunate declined. In 1651 Mitsukuni and his friends began to compile a history of Japan which, when finally completed in 1715, found many interested readers.

While the study of Japanese history brought a new awareness of the emperor's past glories, it also reminded the scholars that there was a world outside and that once Japan had made a brave show in her contacts with it. There was a growing realisation that Japan was in no shape now to defend herself from overseas aggression. Some advocated the new methods of warfare and developing trade contacts with the encroaching powers. Others still clung to the hope that somehow the strangers would be kept out.

In 1862 the Emperor issued an edict that foreigners should be expelled. The Shogun gave a verbal assurance to the alarmed diplomats and traders that nothing would be done. There had already been many murders of 'barbarians' in the streets and two attacks on the British Legation. An Englishman named Richardson was dragged from his horse and hacked to pieces by Satsuma retainers, who considered he should have adopted a more humble attitude as the Lord of Satsuma passed by. Richardson and his female companion had stopped to watch the procession in its quaint medieval splendour. The lady, another hated foreigner, was lucky to escape with her life.

The Choshu clan decided to act on the Emperor's order. In June 1863 their shore guns opened up on American, French and Dutch vessels in the Shimonoseki Straits.

The following year a combined English, American, Dutch and French fleet demolished the Choshu batteries and obtained a pledge that the straits would be left open for foreign ships. The British had already exacted punishment for the murder of Richardson by destroying Kagoshima, the chief city of the Satsuma clan.

With characteristic realism the clans accepted their chastisement at the hands of the foreigners and turned their resentment on the Shogun. He had shown his weakness by leaving outside powers to enforce his policy on his rebellious nobles. At a time when Japan most needed leadership the shogunate was deficient and was allowing the country to 'lose face'.

In 1866 Satsuma and Choshu set aside their traditional rivalry and joined the other western clans of Tosa and Hizen and many of the court nobility in a secret alliance to restore power to the Emperor. Some rich members of the Osaka community, headed by the Mitsui family, gave financial backing.

While the British ambassador, Sir Henry Parkes, was in close

Ho-o-do (Phoenix Hall) of the Byodoin Temple, Nara. Built in 1053, this hall is an outstanding example of religious architecture at a time when the Fujiwara family flourished. The temple is affiliated with the Amida sect of Buddhism.

Great Buddha Statue of Todaiji Temple, Nara. The statue is made of gold and copper; it represents the Rushana Buddha. The building of the temple (741) was a national undertaking during a period known as the 'Golden Age of Buddhism'.

Hirosaki Castle at cherry blossom time. The castle was built around 1600 by the Tsugaru family.

contact with the western clans' progressive young men, the French minister, Leon Roches, gave advice to the Shogun. French instructors trained the Shogun's forces and the Yokosuka iron works and docks, at the entrance to Tokyo bay, were built with French financial assistance. Napoleon III was anxious to increase French prestige abroad by his foreign policy.

In 1867 the Emperor Komei died and was succeeded by his fifteen-year-old son, known as the Emperor Meiji. The Shogun also died and the new Shogun was easily persuaded to restore administrative authority to the throne. However, his supporters were unwilling to accept the disgrace and complete fall from influence that affected them as well as their patron. The ex-Shogun was given no part in the new government and his lands were to be confiscated. Having pinned their faith in his continued patronage, his followers decided to take up arms and attack the Emperor in Kyoto. They were defeated outside the city at the Battle of Toba-Fushimi, named after the two small towns round which the fighting took place.

The foreign diplomats were even more apprehensive for their safety now that the anti-foreign element was in command of the situation. Some Japanese soldiers also felt that the time had come for them to destroy the strangers. They ranged the streets, firing indiscriminately at the 'long noses'. British and American warships landed marines to protect the Europeans. However, in February 1868 an imperial envoy arrived in Kobe, where the foreigners had gathered, to announce that the shogunate was overthrown and to guarantee the safety of the foreign community. The commander of the Japanese troops whose actions had been the cause for alarm was ordered to commit suicide for allowing his men to behave in an undisciplined manner.

The advocates for the modernisation of Japan and contact with the West had won the battle for control of the future. Henceforth the country was to take every opportunity of making up for the two centuries of isolation from the world.

12
Japanese Drama

The origins of Japanese drama, like the roots of her history, are found in the ancient records. The *Kojiki* contains an account of the Sun Goddess, Amaterasu, retiring into a cave in a fit of sulks. To tempt her out another goddess, Uzume, performed an obscene dance, which made the assembled gods laugh heartily. Amaterasu in curiosity finally came out to see what the fun was about, and light was restored in the world.

The story contains the three fundamental aspects of Japanese drama: the supernatural element, found especially in Noh plays; the erotic element, noticeable in Kabuki and some modern theatre; and the dance element.

Religious dances known as *Kagura* were performed before the emperor, the descendant of Amaterasu. These dancers were supposed to be descendants of Uzume, and the dance was a reconstruction of her dance. All Shinto shrines featured a version of the performance, its popularity tending to fluctuate with the popularity of the current emperor. At the Meiji Restoration the dance was revived and can still be seen on festival days in various parts of Japan.

Buddhist temples also had dances of Indian origin. The participants wore masks, examples of which can be seen, though the dances themselves have long since been forgotten. The masks are the forerunners of the masks worn in classical Japanese drama.

A more persistent dance form entered Japan from China in the seventh century. Prince Shotoku was an ardent admirer. By the twelfth century the court was spending so much time on refinements of this form of art, and in composing music for it, that critics of the imperial entourage complained that its members were good for nothing else.

In the country a third type of dance was evolved by peasants: folk dances connected with planting and harvesting rites.

Noh Dance-Drama

At the time of Yoshimitsu, the Ashikaga Shogun at the end of the fourteenth century, Noh dance-drama made its appearance. Some features are reminiscent of Greek theatre. A chorus is used to fill in information and the main action is over before the dance-drama begins. The performance falls into three parts with introduction, development and denouement. The language is archaic and cannot be understood today without a translation into modern speech. Devotees follow on a prepared script.

Noh dramas are performed in a special theatre and performers enter by a passageway to one side of the stage. All female roles are played by males, who speak in their natural voices. Women, demons and ghosts wear masks. Costumes are brilliant. Drums and flute provide the music for the dances, which are slow and stately. Each small movement has a conventional meaning.

Beginning as popular entertainment performed in the temple grounds, Noh was taken over by the aristocracy. In the Tokugawa era it was a punishable offence for anyone below the rank of samurai to witness it.

The themes of Noh are moral and religious. Ghosts are frequently featured. The after-life, the sin of killing, the transience of this world, the evil of lust—all strongly Buddhist ideas —are favourite themes. A programme of five Noh items traditionally includes two comic interludes, the *Kyogen*. These are easily understood. They depict lords mocked by their servants and ordinary men and women in humorous situations.

Japan has always had its minstrels who tell a well-loved story, half-sung, half-recited and accompany themselves on a musical instrument. This form of entertainment is still popular. Stories come from the period of the Gempei wars.

Puppet Theatre

Punch and Judy type puppets were introduced to amuse children in the ninth century. In the sixteenth century they were adapted to illustrate the stories chanted by the minstrels (the *Joruri*).

In 1685 a puppet theatre was established in Osaka and a chanter, a puppet-master and a playwright combined to make an advanced art form out of this children's entertainment. From the eighteenth century the puppets were capable of a variety of movements and required three operators. The dolls are about one-third the size of a man. The manipulators are in full view of the audience. The minor ones wear black, but the major ones stand behind the puppets in ordinary dress. Most people find that they soon cease to be aware of the puppet-operators when they become absorbed in the performance.

Kabuki

The most famous form of Japanese theatre is *Kabuki*. It too had its origins in the dance.

In the sixteenth century a dancing-girl named O-Kuni, claiming to be a performer from the temple girls of the shrine of Izumo, attracted large audiences to her dancing. Her connection with the shrine seems doubtful, but only temple dancers were allowed to collect offerings from their audiences. O-Kuni had to claim to be one of them to make a living. Gradually her fame spread and other performers joined her. Her themes became more elaborate. Drama was beginning to evolve.

O-Kuni is known to have performed before the shogun, but by the seventeenth century the Kabuki theatre deteriorated into a vehicle for immorality. Fearing the effect of pleasure on the samurai, the shogun banned it.

But even persecution and censorship could not stop the growth of popular drama. In 1664 in Osaka was presented the first real play with a continuous story divided into acts. Scenery and curtains followed. The prominence of Osaka in the growth of Kabuki is significant. It represented the new towns dominated by the wealthy merchant class, as opposed to the towns which owed their importance to the courts of the emperor or shogun.

The Shakespeare of Japan is Chikamatsu Monzaemon (1653-1725). His historical plays are not performed regularly today, but his domestic dramas portraying the ordinary life of the period are still popular. Chikamatsu's lovers are always faced with a problem that they can only solve by death. The plots are usually based on the struggle between the hero's human desires and his obligations to the gods, his kinsmen or society.

Since the Kabuki plays were performed throughout the day, from daybreak to sunset, their length made it customary for more than one author to be involved in the writing. The famous *Chushingura* (The Forty-Seven Ronin) was written by three playwrights in 1748. It still attracts large audiences. It is a dramatic masterpiece with plot, sub-plots and development of characters. It calls for a high standard of acting and the leading roles are regarded as the supreme test of an actor's ability. Other Kabuki plays take place during the civil war period when brother fought against brother. The opportunities for a plot involving the dilemma of a samurai torn between natural affection and duty are many.

This form of Japanese drama has rigid conventions which have prevented it from developing into a modern expression of theatrical art. Stylised gestures and dance steps often take the place of acting as understood in the West. Performance of a particular part in the accepted manner requires great skill and the knowledge is often passed down in families, who take pupils to learn their specialty. The dress is conventional and spectacular.

The actors approach the stage by way of the *hanamichi* (the flower way), an elevated platform passing through the audience. It gives extra space and can be used as a bridge or for a road scene. The actor can be seen clearly by the audience and there is something of the intimacy achieved by 'theatre in the round'. The idea was borrowed partly from the Noh stage, but probably owes more to the entry ramp used by sumo wrestlers as they enter the ring. The name 'flower road' comes from the custom of admirers of these unwieldy heroes strewing flowers before them as they advance to meet their opponents.

Some of the stage-craft of Kabuki is quite sophisticated. A revolving stage has been used since 1729, so that a man can be shown walking through a house or a forest. The transfer of the action to the *hanamichi* may allow for a change of scene on the main stage. Trapdoors are used for supernatural appearances.

Curtains may be hung to represent night, mist or other natural conditions, while ingenious changes of dress are affected by the pulling out of a thread which allows the upper garments to be dropped. Underneath a new outfit appears.

Make-up is rather startling, probably due to the fact that in the early days only candles were available for lighting. The face of a good, aristocratic man is painted dead white. The evil, low-ranking, or the brave man is painted in appropriate shades of red. To express more complex characteristics, lines are drawn giving a mask-like appearance. Women's parts are played by men who specialise in female roles.

These ancient dramatic forms hold little appeal for the modern Japanese, who prefers television and cinema. The film industry flourishes in Japan and has reached a high standard, but the stage is neglected. Foreigners find much to interest them in the puppet theatre and in Kabuki, but Noh performers are faced with economic ruin. There is, however, the beginning of a modern drama on Western lines and to please the audiences of American troops the Japanese have found a new talent for lavish musical spectaculars.

Recently Australian dramatists and choreographers have opened up exciting possibilities for a combination of the ancient Japanese ideas with the modern style. Since the dance is so fundamental to Japanese theatre, the interest in the East of Sir Robert Helpmann, the choreographer, holds a promise of a revival in a new form of the scenes that delighted the connoisseurs of Kyoto and the rich merchants of Osaka. Modern Japanese writers and choreographers must surely appear who will follow this trend with their own unique combination of East and West.

13
The Meiji Restoration

Although the faction that had overthrown the Shogunate in 1868 talked of 'restoring' the emperor to power, the term was a misnomer. It was so long since the emperor had wielded any authority that there was nothing to restore. A new system had to be erected. The immediate crisis was covered, however, by a real dig into the past. The Council of State and Ministerial Boards established in the eighth century under the Taiho Laws were adopted and an oligarchy, composed of the able young men of the dominant western clans, took over.

Perhaps it was the study of history again that led the feudal lords to take the momentous step of renouncing their fiefs. The attempt to give the government a strong economic position envisaged by the Taika reforms was accomplished at one blow by the lord's action. The principle that 'there is no soil within the Empire that does not belong to the Emperor' was precisely the one that the Fujiwara had tried to enforce in 646 AD. Their aim, said the daimyo, was to enable the country 'to rank equally with the other nations of the world'.

The government had already reorganised the confiscated Tokugawa lands into prefectures, and it was clear that the fiefs must also undergo some reorganisation. Many of the daimyo hoped that their lands would merely receive a different title and they themselves would remain in control also under a new name. The leaders of the four western clans were finally persuaded that this was not sufficient. A uniform prefectural system must be established. In August 1871 a decree in the name of the emperor stated simply that the 'han' were abolished and prefectures were established in their place. The daimyo ceased to be governors and were ordered to live in Tokyo.

The lords were inspired not only by a natural desire to see Japan a great nation, but also the realisation that unless modernisation was carried out quickly and without bloodshed, the country would suffer the fate of her neighbours.

The end of the nineteenth century was notable for a surge of land-grabbing by the Western powers. England was established in India, the Dutch occupied what is now Indonesia and the French had a foothold in Southeast Asia. China, old fashioned and disorganised, was forced to give in time and time again to demands from the Western powers. Russian possessions were approaching closer to Japan and Russian designs on the islands to the north were well known. In 1860 China ceded to Russia the coast from the Amur River to the Korean frontier and there the Russians built the port of Vladivostok, whose name significantly means 'Ruler of the East'.

A second factor influencing the lords' surrender of their land was that they found themselves financially better off. Burdened with debts as most of them were, it was a relief to cease to be responsible for the expenses of administration and for payment of their samurai. Many of them found that their government bonds as compensation for their losses, enabled them to turn to much more profitable fields of investment.

Finally, the younger samurai of the chief clans were anxious to put their talents to the test and assume the responsibilities of government office.

The ordinary samurai was in many cases left in a very poor economic position, however. His government pension was often too small to meet his needs and in 1876 the situation was made worse by the withdrawal of the annual grant in exchange for a lump sum that was supposed to be the final payment.

Apart from poverty, his pride was hurt by the order to samurai to give up their swords. The new army was to be composed of conscripted men from every walk of life. The government seemed to be going out of its way to belittle the dedicated warrior class by implying that anyone, however humble, could practise the arts of war.

Saigo, one of the leading Satsuma representatives in the government, was very sympathetic towards the samurai. He felt that Japan had made too many concessions and it was time she took a stand. A trading mission to Korea had been rebuffed by the Koreans. Saigo urged that the unemployed samurai should be given an opportunity to restore their self esteem and that of their country by being allowed to lead an attack on their unco-operative neighbour.

Fortunately, a group of statemen who had been overseas trying to negotiate a modification in trade treaties in Japan's favour, returned before Saigo's plan was put into action. They had seen the full might of foreign powers and realised Japan's backwardness more than did Saigo and others who had never left the country. The samurai might be hereditary warriors, but the only fighting experience they had had was in the brawls between competing groups within their own country. The returning negotiators pointed out that, while the Koreans might not be formidable opponents, the Russians could possibly join in and Japan was in no shape to take on a major power.

Saigo was unconvinced and in 1877 he put himself at the head of discontented samurai and challenged the government. The rebellion lasted several months, but in the end the new conscript force proved superior. Everyone realised that the average Japanese, given the training, was in no way inferior to the old warrior class in bravery.

It is notable that Saigo did not see himself as a rebel against the Emperor, but only against his evil advisers. This is perhaps a result of the tendency of the Japanese to mask the real power in the government behind a figurehead. Political assassins are the plague of Japan, but they too always claim to be assisting the emperor against his enemies.

Domestic Reorganisation
The abolition of the feudal system and the reduction of the samurai to a position equal to the other members of the community was eventually carried out, though not without bloodshed. The ex-samurai had to adapt themselves to the changed conditions and became merchants, teachers, scholars and writers.

The fiefs, numbering some three hundred, were replaced by forty-three districts or prefectures. They were controlled by governors appointed by the central authority.

To provide the people with a central figure to attract their loyalty, the emperor's sacred position was emphasised. In 1882 State Shinto was established. The reaction against foreign influences and glorification of anything felt to be typically Japanese had led immediately after the restoration to an attack on Buddhism. Shintoism and Buddhism had become confused and worshippers at a particular shrine were often quite ignorant of the Shinto god it was supposed to honour. Priests of both religions ministered at either temple or shrine without reference to its original sectarian purpose.

In 1868 the Shinto shrines were refurbished and Buddhist statues, gongs and other sacred objects were removed. All Buddhist priests who had become attached to Shinto shrines were to be reordained as Shinto priests or resign. All Shinto priests masquerading as Buddhists were to let their shaven hair grow and return to the native religion.

Between 1868 and 1872 there was a reformation movement that resulted in destruction of Buddhist works of art and sacred texts. Fortunately, the loss was mitigated by priests and worshippers hiding what, despite government decree, they regarded as holy.

In 1882 a Home Department regulation stated that 'from this date on the right of Shinto priests to exercise the function of teachers of religion and morals' was abolished. They were to be government appointees, and were to be paid by the government. They were in fact civil servants, often with no claim to a religious calling at all.

The priesthood certainly stood in need of some kind of reform. To quote a government statement of 1871: 'Even the priestly office of small village shrines has become hereditary. The incomes of the temples have been made family stipends and treated as private property. This widespread practice has continued so long that Shinto priests have come to form a different class from ordinary people and warriors. This does not agree with the present form of government which is the unity of religious affairs and the state.'

In 1868 the first vernacular newspaper was published, but by 1875 the press had become the vehicle for attacks on the government and censorship was imposed.

Obviously if Japan were to catch up with the rest of the

world, her people must be educated rapidly. A democratic system of government also presupposes an educated electorate, but Japanese statesmen were not blind to the centralising and propaganda aspects of a state-run education system. In 1872 compulsory education was established. In view of the difficulties of Japanese writing the literacy rate of the country is astounding. From 1872 Japan became, not only the most literate country in Asia, but one of the most literate countries in the world.

Westernisation was carried out rapidly. In 1869 a telegraph system was begun. In 1870 a start was made on a railway system and in the following years, a postal system and modern style banks were instituted. Tokyo University was founded and a mint opened. In 1873 the Western style calendar was adopted.

A reform of the penal code had to be carried out. The Japanese objected strongly to the protection given to foreign lawbreakers by the treaties which had been signed in the middle of the century, but, not unreasonably, the powers were unwilling to turn their nationals over to be tried under a medieval system of law such as operated in Japan. Japanese envoys sent overseas to try and rectify what was felt to be an insulting clause in a treaty were met everywhere with the same argument. Extraterritorial rights could never be given up until Japanese laws and law courts conformed to modern standards.

It was further pointed out that the nationals of the countries in question were nearly all Christians and as such were still proscribed by Japan's laws and liable to punishment. The edicts against Christians were never formally revoked but they were allowed to lapse, and in 1873 a new penal code and a code of criminal procedure on the French model were set up.

The foundation of the Japanese navy was laid. Great Britain was taken as a guide here, and British naval advisers joined the throng of Westerners who were installing telegraph systems, building railways and generally helping Japan to get up-to-date.

The reformed army was based on the German organisation. Ito Hirobumi of Choshu was much impressed by what he saw of Germany and especially by the talks he had with Bismarck. At his urging, Germans drilled the army, and conscription was introduced—a revolutionary move which helped to break down the old class structure and relieved some of the economic stresses on the agricultural community by removing surplus labour.

The problem of how to bring the Japanese government up-to-date was not settled until the promulgation of the Meiji constitution in 1889. The years from 1868 until that time were spent in trying to find a system that would suit the country. The Charter Oath, issued in the name of the Emperor in 1868 to coincide with the attack on the Shogun's forces outside Yedo, stated: 'Deliberative assemblies shall be established on an extensive scale, and all measures of government shall be decided by public opinion.'

An embryonic parliament was set up the following year with samurai members sent by the daimyo. It was supposed to discuss measures put forward by the ministers, but its main use was a sounding board for the new ideas. Government was actually in the hands of the younger of the leading members of the four main clans. Several other experiments were made, but all proved unsatisfactory.

Meanwhile a growing demand for reforms on democratic lines was given a voice by the *Aikokusha*, formed in Osaka in 1875. Rural landlords, merchants and industrialists and peasants combined to complain that their interests were being neglected. They felt that too much public money was being spent on the development of Hokkaido and other special projects, while the rest of the country was starved for funds. Through the activities of the Aikokusha the government was presented with dozens of petitions asking for an assembly.

A start was made in the establishment of prefectural and urban assemblies in which male subjects over twenty-five and paying a land tax of at least five yen were allowed to vote. Those who paid tax of more than ten yen could stand for election. The assemblies were allowed to discuss matters submitted to them by the local governor, who however could veto their decision. As might be guessed, the stage was set for a struggle for more power.

The two chief political parties in Japan up till 1940 and still existing as two conservative parties under different names, made their first appearance in a simple form about this time. The government formed a party of its own to defend itself from attacks of the other two. Japan was in the curious situation of being a country with three parties and no parliament.

The *Jiyuto* (Liberal Party) was an outgrowth of the Aikokusha. The *Kaishinto* (Constitutional Reform Party) advocated a parliament on the English lines. In their early form the two parties had a brief and stormy existence. Their leaders quarrelled among themselves and the government police gave them little peace. Charges of graft over the sale of government subsidised equipment for the development of Hokkaido, and over trips to Europe paid for by the Mitsui company, discredited the would-be parliamentarians.

In 1881 an imperial rescript promised that a parliament would be established in 1890. The following year a study group, headed by Ito Hirobumi was sent to Europe to make a report on political institutions. Ito was very impressed with lectures he heard by Rudolph Gneist, a German professor of jurisprudence. The argument that the constitution of a country should be based on its historical development appealed strongly to Ito. This meant that the special position of the Emperor must be reflected in any modernisation of the government of Japan.

The study group also visited England, France, Austria and Russia. On their return Ito and nine assistants set about the task of drafting a constitution.

14
The Meiji Constitution

While Ito and his companions debated the form the new constitution was to take, preparatory changes in the government were carried out. It had been decided that there should be an Upper House consisting of peers—as in England and Prussia —but titles of that nature were unknown in Japan. To remedy the deficiency new orders of nobility were instituted. European titles—prince, marquis, count, viscount and baron—were conferred on about five hundred men.

In 1885 a Cabinet replaced the Council of State. Ito, now a peer, was designated Prime Minister. Under him were nine ministers. Inouye Kaoru was at first in the important post of Foreign Secretary, but in 1888 he was replaced by Okuma Shigenobu, a leader of the opposition. The latter had been very vocal against the 'unequal treaties' with the foreign powers. It was a shrewd move to put him in charge of the department faced with the difficult task of negotiating for their removal.

In the same year Ito resigned the office of Prime Minister to Kuroda, and became president of the new Privy Council. This was a key position because the Emperor often attended the meetings and membership was by imperial appointment only. The Privy Council was to have the final say on the constitution when it was finally drafted.

In February 1889, on the anniversary of the founding of his dynasty in 660 BC, the Emperor Meiji handed the new constitution to the Prime Minister, Count Kuroda. Every care was taken to emphasise the fact that the document was a gracious gift from the Emperor to his subjects. It had not been forced on him and was not conceded as a right.

Provisions of the Constitution

The Emperor was described as 'sacred and inviolable'. The right of his line to continue ruling Japan was confirmed in perpetuity. He was to command the army and navy, declare war, make peace and conclude treaties. He could confer rank, titles and civil and military offices. He had the power to open and close parliament and could veto its legislation. When the parliament was not in session, the Emperor could rule by means of ordinances. Even when the houses were in session, he could issue administrative orders 'for the promotion of the welfare of his subjects'. The ministers were responsible to the Emperor alone and not to parliament. It was specified, however, that the ministers could sit in parliament at any time and take part in the debates of either House.

Parliament was to consist of two Houses. The House of Peers consisting of the new nobility—or rather, the old nobility with new titles—had extensive powers. It was able to veto legislation passed up to it from the lower house. The House of Representatives was elected on a property qualification. It met for three months a year and could be recalled by the Emperor if necessary. In between sessions, the Emperor governed by decrees which had to be ratified at the next meeting of parliament.

The seeds of parliamentary control were sown by the provision that parliament must control national expenditure. However, it was largely negated by the statement that if there was any hold-up passing a Budget 'the government shall carry out the Budget of the preceding year'.

The duties of the people were stated to be paying taxes and serving in the armed forces. They were to have freedom of speech and freedom of association. Their homes were not to be entered without their consent and they were granted religious freedom. These privileges could be suspended in time of war or emergency. Freedom of religious belief was granted 'within limits not prejudicial to peace and order and not antagonistic to their duties as subjects'. The rebellious actions and dangerous opinions of Christians and militant Buddhists in the past were not forgotten.

The prominent men of the pre-constitutional period from the Satsuma-Choshu clique formed a shadow government behind the throne. They were known as the *Genro* or elder statesmen, though they were still comparatively young men. They exerted a modifying influence on the government, and although they had no constitutional position their power was considerable.

A more sinister extra-constitutional force was exerted by the Army and Navy. In 1900 it was laid down that the ministers for the Army and Navy should be serving officers of the appropriate force. On the face of it this seems quite a reasonable edict, but in practice it gave the two services control over the government. They could refuse to appoint a minister unless their wishes on policy were met. A Prime Minister was therefore unable to complete his Cabinet without first making sure that he was supported by the Army and Navy heads.

In 1890 the Ministry of Education issued the 'Rescript on Education' to each school. It urged the children of Japan to observe the Confucian principles of obedience towards their teachers, parents and anyone else in authority and to offer themselves 'courageously to the state' if need be. In many schools the Rescript, with a portrait of the Emperor and Empress, was kept in a small shrine and was read out reverently on special days. It was yet another manifestation of the State and Emperor cult of this period.

The Political Parties

Now that parliament had been provided for, the political parties reappeared. They demanded more power for the parliament.

The first elections were not favourable to the government, despite interference exercised through the police and prefectural governors. Neither did they favour exclusively either of the opposition parties, but resulted in the election of large numbers of independents.

Meetings of the Diet tended to be rowdy and the speeches unconstructive. Party leaders were well aware of the weakness of their position as compared with their opposite numbers in the older-established democracies. They saw themselves still subordinated to an oligarchy, since the Satsuma-Choshu clique continued to run the government through its control of Cabinet and possession of ministerial offices. From 1885 to 1918 every Japanese Prime Minister except two (Okuma and Saionji) came

Left: Kabuki warrior portrayed by actor Mitsugoro Bando. Kabuki is Japan's famous dramatic art form. It developed into its present style in the late seventeenth century.

Right: Doll in costume from the Noh play *Hagoromo* or *The Robe of Feathers.* The Noh is a traditional masked drama which was perfected in the fifteenth and sixteenth centuries.

Dress of the period of the Meiji Restoration—a drawing by a Japanese high school student.

Kakizome — the ceremony of 'the first writing of the New Year'. These characters are Kanji; they translate as 'For, see, the wave was calm'.

from one or the other of the two clans. Until a civil service examination system was established in the late eighties, the Satsuma and Choshu clans were a dominating element in the bureaucracy, and even afterwards graduates from Tokyo University were favoured, so that the candidature was still limited.

The only way in which the party leaders could attempt to enforce their demands for a ministry responsible to parliament was through continual opposition to the government on any and every measure put forward. The result was that business was obstructed and the parties looked like a set of useless trouble makers. That this was not the case is shown by the way in which all members co-operated wholeheartedly with the government when the country was at war.

In peacetime, however, the government's reaction to the antics of the opposition was to dissolve the Lower House. During elections, the police, aided by gangs of thugs, intervened quite openly. In 1892 twenty-five people were killed and four hundred injured during elections. Despite this, pro-government supporters were not returned in force, which says much for the independent spirit of the Japanese elector.

In 1898 the two parties succeeded in sinking their differences and their position was then so strong that Okuma and Itagaki, their leaders, were put at the head of the cabinet. The result merely confirmed the oligarchs in their belief that elected members were unfit to rule. The two chief ministers were presented with so many demands for office, and other perquisites of power, by their supporters that they were unable to comply. In fact they had to resign even before the first session of parliament. Two new parties were formed, the *Kenseito* (Constitutional Party) and the *Kenseihonto* (the True Constitutional Party).

The three factors which controlled Japanese politics can be distinguished at this early stage of democratic experiment, and can be traced to the old class system of feudal times.

The *Genro* or Elder Statesmen, such as Ito and Yamagata of Choshu and Matsukata of Satsuma, represented the daimyo. In the early days of the constitution they dominated the Cabinet, the army and the navy, and provided the Prime Ministers. Until their deaths they continued to be the power behind the throne.

The new political parties appear completely foreign in concept, but they came to be associated with the industrial magnates, the former despised merchant class. They inherited also some of the distrust and dislike felt for that class by some sections of the community, who were inclined to blame the wealthy *shonin* for economic distress. The unsavoury use of bribery and grabbing for office, characteristic of the parties, was also felt to be typical.

Finally, the third element of feudal society, the warrior class, showed up in the militaristic cliques which became powerful in the twentieth century.

The lot of the common people under the new, as under the old system, was to obey the laws and pay the taxes.

Ito and Yamagata

The dominant figures in Japanese politics at the turn of the century were Ito and Yamagata. Both were born into samurai households and were of the Choshu clan. Ito, however, had had an opportunity to travel abroad and had realised how much Japan had fallen behind in the years of seclusion. He knew the country needed time for consolidation and renovation. Yamagata, on the other hand, devoted himself to building up the army and felt it was ready to embark on a career of expansion.

Ito, who had initiated the idea of a party cabinet in 1898, in 1900 determined to try the effect of putting himself forward as a party leader. Yamagata resigned and allowed Ito to become prime minister, backed by the Kenseito under a new name *(Seiyukai)*. Unfortunately for the experiment, the combined hostility of Yamagata and the parliamentary opposition, with the inevitable wrangling of his followers, soon compelled Ito to resign. Yamagata and the militarists were now in almost uninterrupted control until after the Great War. As already noted, the provision that the ministers for the Army and Navy should be officers of those organisations, though apparently quite a sensible idea, was used by the militarists to gain control over cabinets and policy.

The parliamentary system and the parties were discredited because of scandals about bribery and scrambles for office.

The Meiji Constitution 75

Strikes and high rice prices were blamed on the tycoons of industry and commerce and it was known that large industrial companies had interests in the parties. They constantly made gifts to party leaders and contributions at election times. Worse still, they were caught issuing bribes to members of parliament whenever any measure affecting their interests was under discussion.

The military on the other hand could point to a successful foreign policy, which boosted national prestige.

15
Foreign Affairs, 1868 to 1905

Immediately after the Meiji Restoration there was a demand that Japan should catch up with the rest of the world in weapons and in territorial aggrandisement. The militarists, so long confined to petty brawls among themselves, urged the government to strengthen the country's position by annexing neighbouring islands and by a more aggressive policy on the mainland.

Russia's interest in Chinese territory near to Japan and in the islands to the north of Japan could not be ignored. Despite her immense size, Russia was deficient in ports which could be used all the year round. The Black Sea ports certainly enjoyed ice-free conditions, but they were useless when foreign control of the Dardanelles could bottle up the Black Sea fleet. An extension of the Russian border to take in part of the Chinese coast would provide good harbours in less vulnerable positions.

In 1861 the Russians landed a force on the island of Tsushima in the Korean Straits on the pretext of protecting their shipping rights. A glance at the map will show its strategic position. However, when the British stepped in and issued an ultimatum, the troops were withdrawn. Japanese action was not necessary then, but during the next ten years they thought it wise to stake their claim to the Kuriles, the Bonin Islands and the Ryukyus. If Japan did not occupy them, obviously Russia might.

Korea

At the end of the nineteenth century the independent kingdom of Korea was torn by factions, which sought to play off the Japanese against the Chinese.

In 1882 the Queen appealed to China for help. In the ensuing troubles the Japanese legation was burned and the staff had to flee. When peace was restored, the Japanese insisted on being allowed to keep troops in Seoul to protect their nationals. Two years later these troops were at loggerheads with the Chinese soldiers in Korea, and the militarists at home were clamouring for action. At the same time the Russians, still anxious to acquire good harbours, offered to train the Korean army in return for the possession of Port Lazarev on the east coast of Korea. Britain, in accordance with her policy of watching Russia's actions in the east, occupied Port Hamilton in the Nan Han islands off the west coast of Korea. This, and the influence of the Chinese at the Korean court, caused the scheme to be dropped.

In 1885 Ito was sent to Peking to complain about the behaviour of Chinese troops in Korea. The result was the Treaty of Tientsin, or the Li-Ito Convention. Both sides agreed to withdraw their troops, to hold discussions on the advisability of the move if they ever thought about intervening in Korea again, and to refuse to allow their officers to train the Korean army.

For nearly ten years the convention served to keep the peace, but Chinese influence was spreading steadily in Korea. Fear of Russian designs in the area were not forgotten. In 1892 the Russians began building the famous Trans-Siberian railway, facing difficulties which place the enterprise among the great engineering feats of the world. Their purpose was to link Vladivostok and St Petersburg (Leningrad), a distance of about four thousand miles. Troops and supplies would be moved by rail from west to east with comparative ease when the line was completed. That the final links would not be made for another fifteen years could not be foreseen by Russia's anxious neighbours.

In 1894, when a revolt broke out in Korea, the Chinese sent troops at the government's request. In accordance with the letter of the Li-Ito Convention, if not with its spirit, they informed the Japanese of their action. The militarists could be restrained no longer. Japanese troops entered Korea, and quickly drove the Chinese out. They swept through the Liaotung peninsula, captured Port Arthur and Weihaiwei and went on to threaten Peking itself.

The Treaty of Shimonoseki of 1895 gave Port Arthur, Dairen,

the southern portion of the Liao-tung peninsula, Formosa and the Pescadores to Japan.

China renounced claims of suzerainty over Korea and agreed to open more ports to trade. Japan was paid a large indemnity.

Russia, France and Germany viewed Japan's progress in China with alarm, and even before the treaty was signed they were rushing round consulting each other on possible action. Britain refused to join in the coercion of Japan, but the other powers agreed to present an ultimatum. With Russian ships in force in Eastern waters, the Japanese had to agree to give up their mainland gains.

In Japan the government's 'cowardly' behaviour was greeted by a wave of suicides and by popular demonstrations. The country was not strong enough to take on the three allies, however. Resentment at the perfidy of the Western nations became more bitter still when, in the following years, all three countries acquired land from China. Russia took possession of the very territory her action had denied to Japan. The lesson seemed clear: only a weak nation need stand by an international agreement or be swayed by ethical considerations. When Japan was strong enough to enforce her will, she too could disregard treaties.

Britain's refusal to join the 'Triple Intervention' against Japan, which followed her renunciation of extra-territorial rights (1894) claimed in the early treaties, made her appear the only power with any sympathy for the newly emerging Asian nation. British statesmen saw Japan as a counterbalance to the ambitions of her rivals in the East.

Japan had gained what should have been a very useful foothold in Korea. The Korean government had agreed to be guided by Japan in carrying out domestic reforms and Lieutenant-General Viscount Miura was sent to give his advice. He proved to be a tyrant. Hordes of Japanese adventurers followed him to seek their fortunes in the land that they regarded as theirs to lord over by right of conquest. Miura did nothing to check their excesses. He even condoned an attack on the Queen's palace and her murder by a mob of Japanese and anti-royalist Koreans. The Japanese government recalled him, but the harm was done. The King of Korea took refuge with the Russians and their influence over the country was restored.

Russia

Japan saw with alarm the growing Russian influence in Korea and China. The Russians helped the Chinese to pay off the indemnity to Japan agreed to in the Treaty of Shimonoseki, and provided capital for railway construction in China. In return Russia was given the right to build a railway across Manchuria to link the line from Vladivostok to the Trans-Siberian line being constructed from the Russian side. The territory through which it had originally been intended to build the final stretch had proved too forbidding and an alternative route had to be found, even if it did mean leaving Russian soil. Although it was supposed to be a joint Russo-Chinese undertaking, the Russians actually had sole control and were allowed to administer not only the railway, but also the land on either side necessary for the maintenance and protection of the line.

A further agreement that Russia would protect China from Japan and would be given Chinese naval bases in wartime was not revealed until the Washington Conference in 1921.

In 1898 Russia acquired the lease of Port Arthur, Dairen and the Liao-tung peninsula. Another rail link from Port Arthur to Harbin, protected by Russian troops, gave Russia a firm foothold in the area.

In the same year the first rumblings of the Boxer Rebellion were heard. It was originally aimed at the Manchu dynasty which had ruled China from 1644. The Dowager Empress and her autocratic government were hated and blamed for China's humiliation at the hands of Japan and the European powers. The government craftily diverted the rebels' hatred towards the foreigners.

In 1900 the Boxers attacked the foreign legations in Peking. Missionaries and traders from all over the country had taken refuge in the legation buildings, and a courageous defence was undertaken for several months before an international force came to their relief.

The Japanese sent a detachment of twelve thousand men which behaved bravely in battle and chivalrously in victory. The

European allies were impressed by their first contact with the Japanese army. Unfortunately, the Japanese too were impressed —by the lawless behaviour of some of the Western troops, especially the Germans. The feeling in Japan that the Europeans were much better at talking about virtuous behaviour than carrying it out, was reinforced.

The Russians took advantage of the situation to pour troops into Manchuria and when the other forces were withdrawn, they maintained theirs, while demanding further concessions from China.

Britain was alarmed at Russia's growing influence in China. An attempt was made to gain German co-operation to persuade the Russians to withdraw. When the Germans refused, Britain turned to Japan and in January 1902 the terms of an Anglo-Japanese Alliance were agreed on in London. Both countries recognised each other's special interests in China, and Britain recognised Japan's special interest in Korea; each would help the other if attacked by more than one power.

America also had been trying to get the powers to co-operate on the question of Russian ambitions in Asia, and in April 1902 Russia finally agreed to withdraw troops from Manchuria in three stages. Playing for time, she merely moved them from one place to another. In 1903 Russian troops were again in Korea.

The Japanese had been ready to recognise Russian interests in Manchuria in return for a free hand in Korea. Ito particularly was of the opinion that a settlement on these lines would be possible, but now the moderates had to give way before the militarists, who could point triumphantly to Russia's belligerent actions.

In February 1904 war was declared. Two days before the Japanese delivered a surprise attack on Port Arthur. Three Russian warships were destroyed at anchor, and on the next day four Russian cruisers were sunk. At the same time a landing had been made by troops who rapidly pushed northwards. By May the fighting was in Manchuria, and Korea was well on the way to becoming a Japanese protectorate.

Port Arthur fell in January 1905. In March Mukden was captured. A Russian fleet from the Baltic completed its lengthy journey to the coast of Japan only to be defeated soundly by Admiral Togo off Tsushima.

The alliance with Britain was a factor in keeping the powers from intervening to aid Russia, and the British government was useful in backing Japan's war loans. The French supported the Russians financially and morally. The Germans too were sympathetic to Russia, but the negotiations for a peace treaty were on the way before the war could spread any further. Through the mediation of President Theodore Roosevelt of the United States the Treaty of Portsmouth (USA) was signed in September 1905.

Japanese influence in Korea was recognised by the Russians. Further, Japan gained a foothold in Manchuria, acquiring the land leased by Russia in the Liaotung Peninsula (Kwantung), the southern section of the Chinese Eastern Railway and Russian-worked coal mines. Both countries agreed to evacuate all troops from Manchuria except those required for protecting their railways.

The southern half of Sakhalin was given to Japan, and also fishing rights off the Siberian coast.

By the Convention of Seoul Korea became a Japanese protectorate and Ito Hirobumi became the first resident-general. China accepted the terms of the treaty of Portsmouth as they affected Manchuria in a separate treaty signed in Peking.

Results of the Russo-Japanese War
President Roosevelt congratulated himself on leaving Japan and Russia to balance the power in the east, but whether the provisions of the Treaty of Portsmouth were instrumental in keeping the peace is open to debate. Like all peace treaties it seemed to hold the seeds of future trouble.

In Japan dissatisfaction was felt that the Russians had not been forced to pay an indemnity, as China had in 1895. There was a general feeling that whenever Japan was gaining at the expense of a Western power the others intervened. Unrest in Tokyo was so great that martial law had to be proclaimed. A public announcement by the Emperor that he was satisfied with the treaty was required before order could be restored. Yamagata also let it be known that Russia was determined to go on fighting rather than submit to an indemnity and Japan's resources were already too strained to undertake any more campaigns.

Japan made substantial gains and established a foothold on the mainland. Her troops' performance had raised the country's prestige in the eyes of the other powers and Britain was so impressed that diplomats immediately began negotiations for strengthening the already existing alliance. Each promised to help the other in case of 'unprovoked attack or aggressive action'.

An important result of the defeat of Russia by Japan was the destruction of the image of 'White Supremacy' over Asia. Nationalist movements were encouraged by the thought that what one Asian country could do, others could do likewise.

Russia's interest was turned once more westward. Foiled in the attempt to extend her influence in the East, Russia turned to look with added interest at what was happening in the Balkan area on the other side of her vast territory. Events there were rapidly building up for the crisis which began the First World War (1914-1918).

Japan's relations with Korea were never happy after 1905. Ito was assassinated by a Korean fanatic in 1909 shortly after he had resigned as resident-general. Japan used this as a pretence for taking over the country completely. For the next thirty-five years Korea was ruled by the Japanese military whom the whole country grew to hate. Koreans were treated as subordinates in every way—educationally and in business opportunities. Even as labourers they were paid less than the Japanese doing the same job. The country became a huge military base and its products were developed with the sole aim of strengthening Japan. Good roads, good railways and spreading industry brought little benefit to the indigenous population.

Finally, perhaps the most important result of the Russo-Japanese War was the effect of the defeat of Russia on the Japanese themselves. Since returning to the community of nations they had indulged in two wars. The one against China had resulted in a quick defeat of a country which had always been regarded with respect, both for its size and its superior civilisation. In the second war Japan had in a few months soundly beaten a major European power.

The self-confidence engendered was unfortunate. The militarists were bolstered in their claims to know what was best for the country. The complaints of the moderates that the economy could not stand an extended campaign was branded as defeatist pessimism. The mood of the country was all set for the military adventure of 1941.

16
Relations with America

After 1905 the European powers began to reassess their relationship with each other in view of the deteriorating political situation. Instead of Russia being the threat to peace, it was becoming clear that Germany was likely to be cast in the role of aggressor. Instead of the region of conflict being hundreds of miles from the European homelands, it threatened to be right on their doorsteps.

Britain, Russia and France decided the time had come to settle their differences in the East, and in 1907 these were brought to a satisfactory conclusion. Japan, as Britain's ally, was able at last to carry out Ito's policy of an agreement with Russia. Japan recognised Mongolia as Russia's sphere of interest, while Russia recognised Japan's claim over Manchuria and Korea. At the same time France and Japan agreed to maintain the territorial integrity of China. The Russo-Japanese division of China's territory was kept secret, so that the agreement with France did not appear in 1907 quite so two-faced as it looks today.

Japan's relations with America were not so happy. America was acquiring islands in the Pacific which brought her close to Japan. These were territories which, but for the years of seclusion, might have become Japanese possessions. We have noted the activity of Japanese traders in the area up to the seventeenth century, and Japanese threats against the Spaniards in the Philippines, which were taken seriously by the Europeans. By the twentieth century the opportunity for expansion was gone. America had stepped into the power vacuum left by Spain's decline and annexed Hawaii and the Philippines in 1898.

California had been acquired from Mexico in 1848. The former Spanish settlement had no sooner been taken over than gold was discovered, attracting a rush of immigrants.

With their return to communion with the rest of the world, the Japanese began to look again towards eastward expansion. Disputes over immigrants into Hawaii began soon after American annexation, but real trouble threatened when Californian labour groups began complaining about the influx of cheap labour. The area had already received a sizeable Asiatic population in the shape of Chinese immigrants during the gold rush. Many Californians felt that people from an alien culture were likely to engulf them if unrestricted immigration were to continue.

Japanese resentment rose high when in 1906 San Francisco officially barred Japanese children from public schools. President Theodore Roosevelt succeeded in persuading the state to rescind the order and in return undertook to limit immigration. The Japanese government proved co-operative, and restrictions were enforced by a 'gentleman's agreement' between the two countries.

In 1912 an American group planned to sell a large area of land in California to a Japanese syndicate. The government stepped in and forbade the deal on the grounds that the territory in question was situated in a position to command the western end of the Panama Canal and therefore was of strategic importance to America. Alarm at Japanese interest in the west coast was raised again and the result was the passage of the California Alien Land Law Act, 1913. It was made impossible for land to be held by 'aliens ineligible for American citizenship'.

The old question of racial discrimination was brought to the fore again. Many Japanese settlers were now successful fruit farmers and had adopted California as their home. The new land law branded them afresh as aliens without citizenship and without any chance of becoming owners of their farms.

At the same time that America was cutting down on Japanese ventures in American controlled territory she was claiming that the Chinese mainland should be left open for trade (the Open Door Policy). The extension of Japanese interest in China was viewed with distrust by the United States. Once again Japan felt that the Western powers had one policy for themselves and another for the Japanese.

In 1909 an Anglo-American group received permission from

the Chinese government to construct a railway through Manchuria. The American Secretary of State, C. Knox, proposed that all railway communications in Manchuria should be internationally controlled until China could purchase them with the aid of foreign loans. Russia and Japan saw this as a threat to their interests and agreed to co-operate with each other to protect their investments.

By 1912 Britain was seriously alarmed at the growing divergence of views between Japan and America. In view of the attitude of Germany in Europe, where it was clear that a war atmosphere was thickening, Britain felt that her two potential allies must be kept from each other's throats if possible. Britain's treaty with Japan might have involved her if war were to break out between the two. Japan, however, agreed not to hold Britain to her obligations if Japan's enemy was a power with which Britain had an agreement, and an Anglo-American treaty in 1914 provided that all disputes between the two countries should be referred to a special investigation committee.

The outbreak of the First World War shelved, for the time being, the disagreements affecting the East. Japan and America found themselves in sympathy with the same allies for the next four years.

17
Japan as a Major World Power

Internal Development after 1867

The overthrow of the Tokugawas and the antiquated system they had administered made the lack of a fiscal policy in Japan even more obvious. The old system had been gradually creaking to a standstill, but as it was wiped out at one blow, some temporary stopgap had to be found.

In the crisis the Emperor called on the Mitsui family to provide him with a loan. Together with other great merchant families like Ono and Shimada, the Mitsuis gave large sums to the new government and the crisis was overcome. These early favours were not forgotten and mutual co-operation between merchants and government continued during the succeeding years.

Family concerns like Mitsui were the foundation for what are known as the *Zaibatsu*—family trading combines. The Mitsui relations all worked for the company, but were subject to rigid discipline, learning the business from the lowest level. They and other great merchants began as employees of daimyo or of the Imperial Court. Their duty was to receive the dues from the estates of their master and administer them to his advantage. Naturally the position was one which enabled the holders to acquire wealth and influence, as well as skill in business matters.

The ready capital supplied by the merchants was not enough to supply all the Emperor's needs as he assumed control over the country. To the gold and silver coins, in various stages of debasement, and private notes issued by daimyo, was added a government note issue. As more cash was needed more was printed.

The daimyo's surrendered fiefs were now in theory the possession of the state, but the country was so disorganised that revenues were not immediately forthcoming. The Emperor had undertaken to pay pensions to the daimyo and samurai and these were met by further note issues. Wholesale inflation should have followed, but the notes proved to be a useful means of exchange and were based on the real wealth of the country. The note issue, uncontrolled as it was, apparently never outstripped the actual demand for a means of exchange to facilitate the country's trade. The government, when it realised the danger, gradually exchanged notes for government bonds carrying six per cent interest. The pension payments to the samurai and daimyo were also changed into bonds in 1877.

A great number of banks were set up in the early years of the restoration. Some were begun by merchants, but others were established by groups of samurai and nobles, who used their new government-supplied capital for this purpose. Most of the latter were failures because of the lack of business experience of the directors. In 1882 the banking system received a thorough overhaul. The Bank of Japan was founded to act as central banker and the Mitsui concern, the Yokohama Specie Bank, was entrusted with foreign exchange. Its president and vice-president had to be authorised by the Minister of Finance. Industry and agriculture were to be assisted to develop by special banks which could lend on types of security that would never be accepted in more highly capitalised countries. A Post Office Savings Bank catered for the small savings that were typical of Japan. The accumulated funds could be used for development of the country through government agencies.

The government was determined to modernise as quickly as possible. It began with cotton mills, silk mills, pottery and similar businesses. When established, these concerns were made over to private enterprise. Steps were taken to develop Hokkaido. Emigration to the northern island was encouraged and a brewery and sugar factory were set up by the government in Sapporo. The chief exports were tea and silk. Fortunately for the Japanese farmer the opening up of Japan coincided with a disease of silk worms in the European silk-producing countries. Both silk and silk worms were therefore in demand.

Distress was inevitable as people who had been employed in catering for the needs of feudal retainers sought other means of support. Gradually they were absorbed into the new factories or returned to help with the silk production on their parental farms. The army also absorbed large numbers of young men.

In 1881 Prince Matsukata, Minister of Finance, initiated regular annual budgets estimating expenses and providing for their payment. Taxes on sake and tobacco supplemented the land tax and a sinking fund was established to meet the huge national debt the government had contracted by the issue of bonds. At the same time advantageous sales of government-established factories and the reduction of rates of interest on government loans increased the balance slightly in the government's favour.

The Zaibatsu, which had earlier come to the government's aid, now received their reward. The big four came to be Mitsui, Mitsubishi, Sumitomo and Yasuda.

We have already seen the origins of the Mitsui wealth. Mitsubishi was founded by an ex-samurai who first acquired several ships from his daimyo as the feudal system was liquidated. These he made available to the government for the transport of troops to Formosa in 1874. When the government bought ships from overseas for the same purpose they were entrusted to the Mitsubishi concern. In 1875 Mitsubishi took over Kaiso Kaisha, a transport company originally set up with state money. With thirty-seven ships now in commission, the ex-samurai began regular services to Shanghai. These were later extended to Hong Kong and Vladivostok.

After Mitsubishi ships had again been employed by the government to transport the troops to put down the Satsuma rebellion, it was only reasonable that the firm should be rewarded for its loyalty by further assistance. When the famous shipping line Nippon Yusen Kaisha (NYK) was formed, the government guaranteed dividends of eight per cent on its stock. The wealth of Japan was pledged to support this Mitsubishi enterprise.

During the same period firms such as Hitachi and Kawasaki were beginning in a small way and with similar encouragement from the government. This is a special feature of the development of Japanese industry.

Also noteworthy is the paternal interest industry takes in its employees. Most big firms provide hostels for their workers to live in, a legacy from the times when the factories were being established in the towns, while the workers had to be brought from the country. It must be admitted that in the early days the factory manager actually bought young people from their parents and took them to work in the towns. Nowadays, while this unpleasant aspect has gone, the provision of living quarters, recreation facilities, dowries, etc., are still features of Japanese staff amenities. Bonuses to employees are given at the traditional gift-giving times of New Year and O-Bon.

Up to 1914 Japan remained predominantly agricultural, but there was a gradual rise in urbanisation. Rice production increased, but some products, such as cotton, ceased to be a profitable crop for the Japanese farmer. He could not compete with the cheaper products of India. Cotton and woollen milling, with imported raw materials, increased rapidly. There was an ample supply of cheap labour available and a tradition of textile work derived from the production of silk in cottage industry.

Heavy industry was slow to develop. The government had to supply the capital and at first the state was the only market. Iron ore and coking coal had to be imported. There was an increase in coal mining in Kyushu and Hokkaido, but the industry was inefficient. The only mineral Japan had in any quantity proved to be copper. By 1914 Japan was the world's second largest copper exporter. The industry was in the hands of the Zaibatsu.

Japanese industry benefited from the boom of the years of the First World War. Textiles and shipping were in demand. Japan acquired a large trade surplus, but her inexperienced financial ministers did not invest it wisely. The wage and price structure was upset. Food became very expensive. In 1919 the country was the scene of rice riots. The next ten years were very difficult economically for Japan.

In 1923 Tokyo was destroyed by a severe earthquake—which, however, created employment by the necessity to replace goods and houses.

Uncontrolled lending by the banks caused many to fail. In 1927 the government had to resort to the time-honoured Japanese expedient of a moratorium on debts. The central bank rescued some from disaster by loans, but on the whole the crisis was beneficial to the economy. It got rid of the most inefficient industries and banks.

The slump in America and the West hit Japanese farmers because the demand for silk fell. Heavy rice crops led to a fall in price for the farmer's staple crop.

Industrial Development after the World Depression

The government set up an Industrial Rationalisation Bureau as part of the Ministry of Commerce and Industry. Its object was to increase efficiency. A feature of Japanese industry is the close relationship between different branches of manufacturing. The assistance given by the government enables it to dictate terms to the recipients.

During this period the militarists were becoming increasingly powerful. The Zaibatsu were associated with the civilian political parties. Throughout the country they were unpopular because they were blamed for the depression. The industrialists were obviously still prosperous and it was felt that they were only interested in their own well-being.

In 1932 Baron Dan, the chief executive of Mitsui, was assassinated by a group of young officers. Mitsui tried hard to recover its public image. Officials who had been responsible for business policy were dismissed. The public was allowed to buy large holdings of shares in Mitsui companies. Immense sums of money were given for relief of distress and to establish social services. In 1936, after a further wave of army-sponsored assassinations, prominant members of the Mitsui family withdrew from active participation in the business. They tried to show their patriotism by investing in industries of national importance, such as the production of synthetic oil.

The militarists encouraged the growth of the 'new Zaibatsu'—new industrial groups. Co-operatives of peasants attempted to exclude the big firms from rural trade. Heavy industry was fostered by the militarists, who aimed at controlling everything which could contribute to their drive for power. Manchukuo was administered by the army through the Manchukuo (as the Japanese called Manchuria) Railway Company. Metallurgical and chemical plants were developed there.

In 1934 the Japanese Iron Manufacturing Company was formed by the amalgamation of six large private concerns and the state's Yawata works. Nearly all pig iron production and half the steel production thus came under the direct control of the government, which, by this time, was controlled by the militarists.

Oil production was brought under government control and all firms dealing in oil, including foreign ones, had to keep six months' supply in store. The government also assumed control over shipping companies and encouraged the building of new, fast ships. In fact, the country was completely geared for war. Goods for domestic consumption produced by the Zaibatsu would not be needed by a people mobilised as the Japanese were to be for the next few years.

18
Japanese Politics, 1918 to 1945

Up to the 1914-18 war Japan was controlled by the leaders of the old clan system, notably the Satsuma-Choshu group. After the war three main factions began to emerge.

On the extreme right were the ultra-nationalists, the militarists. They regarded assassination as a legitimate means of achieving their aims. Parliamentary government was to them only an instrument through which they could enforce their policy of expanding Japanese influence.

On the left was a growing body of socialist and later communist thought. Katayama Sen led the Japanese socialists as early as 1896. He was attracted by the Communist International ideal of unity of all workers. He became one of the first Japanese communists and was to end his life in exile in Moscow. Kotoku Denjiro was another leftist, who led anarchist riots in Tokyo in 1908. He was executed in 1911 for an alleged plot to assassinate the Emperor.

Like young intellectuals all over the world, the Japanese students especially were attracted to the Marxist ideas which culminated in the Russian revolution of 1917. The movement was hampered by the fact that Russia was Japan's rival during the early part of the century, and by the repugnance that most Japanese felt for attacks on reigning monarchs, personified for them by their emperor.

In 1912 a Friendly Society was established to mediate between employers and workers for better conditions. An embryo trade union, it was the first move for this kind of collective bargaining in industrial relationships to be seen in Japan.

The Seiyukai and Minseito parties represent the 'Liberal' element in Japanese politics. As noted previously, they tended to be associated with the Zaibatsu. Doubts were felt about their integrity because of the support they received from wealthy quarters and the interest their numbers seemed to have in 'feathering their own nests'.

From 1918 to 1932 the parties were in theory in control of the government and party leaders were Prime Ministers, but the army was able to dictate policy by refusing to appoint an army minister until certain conditions were met. The military clique was in a key position.

Hired bullies were increasingly employed by all kinds of organisations to smash opposition. Labour unions and strike action were broken up by this means and liberal newspaper offices were attacked. Fights among members of parliament were common. In 1921 the Prime Minister, Hara, was assassinated.

The great Tokyo earthquake of 1923 provided an excuse for a wholesale attack on socialist and left-wing elements. It was said that they and Korean nationalists had planned to take advantage of the confusion to set up a revolutionary government. Hundreds of people were arrested and many were murdered in the streets or by the police.

What appeared to be a liberal movement in which, in 1925, the vote was given to all males over twenty-five, was balanced by a determined drive against the left wing. A Peace Preservation Law provided that those who formed, or joined, societies aimed at altering the constitution or form of government, or repudiating private ownership of property, could be imprisoned for up to ten years. A campaign against 'dangerous thoughts' was conducted by the reactionary minister for education, Dr Okada. Students were arrested and the study of Marx was forbidden. Hundreds were imprisoned under the Peace Preservation Law.

Unlike the Facist movements in Germany and Italy, where big business allied itself with the military, the Japanese middle class found itself opposed to the warlike trend followed by some army elements. Their interests were not favoured by warfare. They preferred to cater for a growing overseas and domestic consumer market. Indeed, as we have seen, war industry was already out of their hands.

In 1927 Hamaguchi organised a Minseito party cabinet backed by Mitsubishi. They tried to cut back military expenditure and pursue a conciliatory policy in foreign affairs. Unfortunately, 1927 was a year of economic crisis. The business leaders were held responsible and were the objects of a growing hatred.

The House of Peers backed the militarists in their demands for an active policy in China. The government could not enforce its alternative. Control of events was for all practical purposes out of its hands. In Manchukuo the army ran heavy industry and was practically independent of interference from home. After an almost successful attempt to assassinate Hamaguchi in 1930 (he died the following year), the chief activity of the Foreign Office was trying to justify the actions of the army and cover up the fact that the Foreign Minister was no better informed about what was going on than was any member of the public.

The Militarists in Power, 1923-1945

With the death of Marshal Yamagata in 1922 the domination of the Choshu clan over the army ended. He had insisted on keeping the army out of politics, but with the younger army officers coming from other clans and often without a samurai background, a new attitude began to develop. While they cultivated the courage and arrogance of the old officer class, the younger men discarded the finer points of the code of Bushido. They inherited the samurai dislike of, and contempt for, the merchant class. The new army was recruited mainly from the country. The sons of farmers readily joined in the dislike for the wealthy urban groups, especially when the effects of world depression hit the rural areas.

The 1914-18 war was followed by a reduction in the army, which threw numbers of officers out of employment and added to resentment. The ex-serviceman's society fostered these feelings and brought soldiers and ex-soldiers together to air their grievances.

In 1926 the old Emperor Taisho died and was succeeded by Hirohito, who had been Regent for the past five years while his father had been mentally unbalanced. The new regime, known as the 'Showa Era', gave its name to a movement, the Showa Restoration, which was aimed at overthrowing the Meiji constitution.

In May of 1932 a group of young naval officers and army cadets invaded the house of Premier Inukai and killed him. Earlier in the same year Inouye, a former Finance Minister, and Baron Dan, chief director of Mitsui, had been murdered by an ultra-nationalist group known as 'the League of Blood'.

This marked the end of the attempt to govern through parties. The next premier was Admiral Saito, but the most powerful man in the cabinet was General Araki, Minister of War. He was leader of what was known as 'the Imperial Way School'. The army was actually split between two factions. One wished to advance against Russia, the other against China. However, both were united in opposition to what they considered to be the overwhelming influence of the Zaibatsu, and in their desire to lead Japan to greatness through military power.

In 1934 a plot to remove the Saito Cabinet by bombing them from the air was discovered in time, but the next Cabinet, that of Admiral Okada, was not so fortunate. In February 1936 two infantry regiments formed a number of groups and simultaneously attacked the homes of the Prime Minister and other public figures. Admiral Okada escaped with his life, but ex-premiers Saito and Takahashi were killed. Admiral Suzuki was left for dead, but survived to become premier nine years later.

The rebels occupied the centre of Tokyo and declared that Japan's ills were due to the *Genro* (the Elder Statesmen), the Zaibatsu and the political parties. They claimed to be acting on behalf of the Emperor. He, however, publicly disowned them and ordered the loyal section of the army to put down the rebellion. Seeing that they received no support from the high-ranking officers, the rebels put down their arms without a fight. The ringleaders were executed. Former assassins had received much public sympathy and been let off with prison sentences, but the Emperor's firm stand and the general alarm felt at the anarchy within Tokyo lost any backing the rebels might have hoped for.

The army continued to dictate the composition of the government through its control of the appointment to the head of the War Ministry, but general elections showed that the people were opposed to the military. A compromise was finally reached with the appointment in 1937 of Prince Konoye as Prime

Minister. He was liked by everyone and, as a member of the ancient Fujiwara line, he could be expected to be above politics and petty considerations. Unfortunately, he had not the strength of will to carry out a firm line of policy and was too easily swayed by arguments.

Japan was hard hit by the depression which had overtaken world markets. Konoye called together prefectural governors, leading industrialists and financiers for a consultation on economic affairs. He realised that only restored prosperity could block the ambitions of the military clique. Through the Bank of Japan measures were taken to prevent the yen from falling in value on the foreign exchange market. As already noted, the leaders of the Zaibatsu took steps to relieve their unpopularity.

There was, however, no holding the army, even though the Emperor himself expressed disapproval of its actions. The cabinet was informed that reinforcements were already on their way to China; and what had begun as a defensive occupation of Manchuria against the aggressive moves of Russia ended as a full scale war against a neighbouring country.

Konoye's 1938 cabinet reflected the new state of affairs. General Araki became Minister for Education and directed an increase of ultra-patriotic and militaristic teaching in the schools. General Ugaki became Foreign Minister. Ikeda, a financier and former head of Mitsui, became Finance Minister. The leaders of the radical movement were consigned to prison and the country was organised for war.

19
Foreign Affairs, 1914-1939

When the First World War broke out in 1914, Japan had treaty agreements with Britain which bound her to enter the war if British possessions in the East were attacked. The militarists saw an opportunity to gain coveted Chinese territory while pleasing their allies. Eleven days after the declaration of war in Europe Japan demanded the surrender of Tsingtao, which was held by Germany. It was further stipulated that all German warships should withdraw from Chinese waters. When the Germans did not comply, an attack was launched and a combined Japanese-English force soon took the town.

Although this was the only land action fought in the Great War by the Japanese (in which, incidentally, the British troops were under Japanese command), the Japanese navy proved very useful in defending Eastern waters, escorting Australian and New Zealand troopships and taking over German islands in the Pacific area. Later they were equally ready to do escort service in the Mediterranean at the height of the submarine menace.

From the beginning of the nineteenth century China had been disintegrating and falling piecemeal into the hands of the powers. The country was almost as backward as Japan had been in 1853, but whereas the latter country immediately bent every effort towards catching up with the rest of the world, the Manchu rulers of China continued to be told by their sycophants, and to believe, that they were the rulers of the world. That the 'foreign devils' or 'long noses' were not rebellious subjects, but powerful invaders, was slowly brought home to the Chinese as treaty ports were opened to trade and pieces of territory were occupied by Russians, English, French, Germans and even by that former humble admirer, Japan. The Chinese felt keenly the success of the 'dwarf bandits', as they called the Japanese. It seemed unforgivable that a people who had learned so much from them should now be allying themselves with the hated foreigners.

During the twentieth century the Manchu dynasty came to an end and chaos reigned. The country was divided among warlords, who fought each other, preyed on travellers and oppressed the peasants. In 1911 Sun Yat-sen founded the Kuomintang. A republic was proclaimed and Sun began the task of unifying China once more. When he died in 1925 General Chiang Kai-shek assumed the leadership. At first he was helped by the new Communist Party, but soon he decided that the Communists were a worse threat to his plans than were the warlords.

Such was the state of Japan's neighbour, a neighbour which possessed the raw materials and the undeveloped land that Japan lacked. The involvement of the European powers in warfare seemed a heaven-sent opportunity. Less than six months after taking over the German possessions in Shantung the Japanese minister called on the Chinese President and presented him with the 'Twenty-One Demands' (1915).

In addition to keeping German rights in Shantung, the Japanese wanted commercial freedom in south Manchuria and a ninety-nine year lease on mines there. They demanded a half interest in the Hupei iron mines and steel mills and in a Szechuan colliery, and an assurance that no part of China's coast should be ceded to any power.

The Chinese President told Paul S. Reinsch, the American Ambassador: 'I am convinced that the Japanese have a definite and far-reaching plan for using the European crisis to further an attempt to lay the foundations of control over China. In this control of Shantung through the possession of the port and the railway is to be the foundation stone. Their policy was made quite clear through the threatened occupation of the entire Shantung railway, which goes far beyond anything the Germans ever attempted in Shantung Province. It will bring the Japanese military forces to the very heart of China.'

In May 1915 Japan presented a modified form of the 'Twenty-One Demands', accompanied by an ultimatum. The portion omitted is significant because it shows the Japanese estimate of

China's weakness. This section originally had ordered China to accept Japanese political, financial and military advisors. Japanese citizens were to be given full rights to hold land in China and to construct railways in the southern part of the country.

The Chinese had no alternative. They had to accept the amended Demands. The United States was the only outside power to register any opposition, though the protest was feeble. America was anxious to bolster up the new Republic of China. The so-called 'Open Door Policy' advocated unrestricted trade facilities for all as far as China was concerned.

When in 1917 both America and China entered the First World War on the side of the allies, it was obvious that the Chinese would now be able to state their case, with American support, at the peace treaty conventions. At the ensuing Versailles Conference Britain and France backed Japan as they had earlier promised they would. Despite Chinese claims that their participation in the war entitled them to a favourable hearing, including a revision of their agreements under duress with Japan, they were forced to concede 'economic privileges granted to Germany' to the Japanese. Though the Shantung Peninsula was to be handed back 'in full sovereignty' to China, the Chinese refused to sign the treaty as they were not satisfied with the terms. Therefore, the Japanese continued their military occupation of Shantung, which suited their plans very well.

At the same time the Japanese were able to establish troops in Manchuria, taking advantage of the confused situation in Siberia. The Communist revolution in Russia had thrown the whole country into confusion. The Allies were anxious to keep the Russians in the war on their side and troops were sent to Siberia to protect foreign nationals and military stores belonging to the Allies. For a while it looked as though the anti-communist forces would get the upper hand, but they were eventually defeated. The powers were not prepared to commit themselves to a full scale war against the Bolsheviks, and they contented themselves with protecting their own interests. The Japanese argued that their interests in Manchuria were so vital that troops would have to be maintained there.

In addition to these two footholds on the Chinese mainland,

Japan emerged from the First World War with a mandate over the former German colonies in the Pacific—the Caroline, Mariana and Pelew Islands.

The Washington Conference

In 1921-22 a conference was held in Washington to try to clear up problems left over from the war. The first was the problem of disarmament. The second was the problem of the Asian mainland. Of the eight powers invited to attend, only two— China and Japan—were truly representative of Asia, though France, Great Britain, the Netherlands, Portugal and the United States had territorial interests there. Italy and Belgium were also represented, but Russia was still not regarded as sufficiently stable to receive an invitation.

The question of a reduction of war shipping was settled by a naval treaty, limiting the navies of the powers involved in the following proportions: Britain 5, USA 5, Japan 3, France 1.75, Italy 1.75.

Fortifications and naval bases were allowed on certain islands:

America was allowed the Aleutians, Guam, Pago Pago and the Philippines.

Great Britain was allowed Hong Kong, Singapore and Pacific Islands held by Britain.

Japan was allowed the Kuriles, the Bonins, Amami-Oshima, the Ryukyus, Formosa and the Pescadores.

The Anglo-Japanese alliance was replaced by a Four Power Treaty. England, Japan, America and France agreed to consult each other on Pacific problems should they arise. They would respect each others' rights in the region.

On the question of China, nine powers signed an agreement to respect China's sovereignty and to give the country every assistance in developing a stable government and in increasing her trade. The Chinese and Japanese delegations to the Washington Conference also discussed the question of Shantung and Manchuria. The Japanese agreed to withdraw their troops from Shantung, but maintained control over the Tsinan-Tsingtao railway in that province. They refused to be dislodged from their foothold in Manchuria.

A final problem left over from 1918 was the question of

Japanese troops in Siberia. They were withdrawn from the mainland in 1922, but it was not until 1925 that they were withdrawn from Sakhalin. The Soviet Union in return granted concessions to the Japanese for the exploitation of mineral and forest resources in Sakhalin. In later years they were gradually but decisively squeezed out.

The militarists had received a few setbacks at the Washington Conference, but they had a strong base in Manchuria, where they worked in close co-operation with the local warlord. The 'Kwantung Army', as the troops of the mainland were called, was a law unto itself. While the rest of China was torn by civil strife, Manchuria was peacefully exploited by the Japanese. While the members of the government at home worked to improve relations with foreign powers, the Kwantung Army pursued an independent course.

In 1928 the warlord, Chang, was murdered. Since 1945 it has been revealed that the deed was the work of Japanese army officers. At the time it was thought to have been ordered by the Japanese government. The whole episode is an illustration of the independence of the Manchurian troops and the lack of discipline that prevailed. It is now known that Baron Tanaka, the Prime Minister, ordered the punishment of the offending officers, but army privilege overruled him.

Chang's son, however, was not so easily disposed of. 'The Young Marshal', as he was called, became bitterly anti-Japanese. In 1929 he threw in his lot with Chiang Kai-shek and the Chinese set themselves the task of freeing Manchuria from foreign influence.

They turned first against the Russians. The Chinese Eastern railway in north Manchuria was taken over. The Russians immediately sent troops who quickly disposed of the Chinese. The Japanese army viewed the Russian success with interest and alarm. Their own relationship with the Chinese was rapidly deteriorating. Japanese and Chinese army officers insulted each other openly in the streets and restaurants, Korean and Japanese settlers in China were continually at loggerheads with the local people.

Finally, a Captain Nakamura, who was supposed to be touring western Manchuria as an agricultural expert, was shot by the

Chinese as a spy. Japanese officers now demanded action. They probably engineered for themselves what is known as the 'Manchuria Incident' of September 1931, when an explosion on the south Manchuria railway was investigated by Japanese troops, who claimed then that they were fired on by Chinese. With this as an excuse, the Japanese army turned out in force and captured Mukden.

The Lytton Commission
Lord Lytton, leader of the League of Nations Commission of Enquiry into the incident, reported that it appeared that the Chinese had 'no plan for attacking the Japanese troops'. Nevertheless, the Japanese went on to seize the whole of Manchuria. In 1932 they proclaimed its independence. Henry Pu Yi, who had abdicated from the throne of China in 1912, was installed as Emperor, the puppet of Japan, in February 1932.

The Lytton Commission in 1933 recommended that Manchuria should be restored to China, and the League of Nations refused to recognise the new government. Japan immediately left the League, which she had joined with pride at being accepted as a member of the community of nations. Japanese anger at the League was even more pronounced because Russian action in increasing control over Mongolia and Sinkiang had passed without protest. Russia was not even a member of the League. Outer Mongolia had already become communist and Japan was afraid that the rest of Mongolia, perhaps even Manchuria, would follow suit.

Fortunately for Japan, Chiang Kai-shek refused to co-operate with the Chinese Communists and even fought against them instead of against the Japanese.

In 1934, however, the situation was beginning to change. Russia joined the League of Nations. After sixteen years of hostility and suspicion from European powers and the United States—who viewed the excesses of the Russian revolution with dismay—the Soviet Union was anxious to restore normal relations. During the 1930's the growth of fascism in Europe seemed more alarming than communism. Japan's behaviour was beginning to look suspiciously like fascism too.

In 1936 Chiang Kai-shek was kidnapped by the Young Marshal

and forced to agree to join the Chinese Communists in fighting the Japanese. The Russians, too, had been urging the two groups to unite against the common enemy. Although the type of government in Russia had changed completely from a monarchy to the 'dictatorship of the proletariat', the foreign policy of each proved to be remarkably similar. The communist leaders were just as anxious to extend and protect their interests in the extreme south-east of their immense territory as had been the Czars. Despite the much publicised international aspect of the new dogma, their encouragement of a united stand against Japan was also in their own national interests.

So far Japanese efforts had been mainly in Manchuria, but, faced with a united enemy, they began operations in the heart of China. Troops had been stationed around Peking to protect Japanese nationals ever since the Boxer Rebellion. They now indulged in 'night manoevres', which ended in a clash with Chinese forces. Known as the 'Marco Polo Bridge Incident', this was an excuse for a full scale attack on Peking with ground and air support. The Chinese had no choice but to abandon the city.

In 1937 Japanese armies captured Shanghai after heavy fighting. Their troops were successful all over north China. At the end of the year they had pushed up the Yangtze and captured Nanking. Chiang Kai-shek refused to surrender. He moved his capital to Chungking and, with the aid of military supplies from Russia, England and America, continued the fight.

The Japanese blocked the coast, extended their power over most of north China and bombed Chinese cities unmercifully. Between 1937 and 1939 the Burma Road was built to enable Allied supplies to reach the Chinese armies. In 1939 too the Russians and Japanese came into conflict in north China, ostensibly supporting their respective allies, Manchuria and Mongolia.

In the same year the Japanese received a shock from their friends the Germans, when the Russo-German Pact was made public. Once more Japan was given a lesson in diplomacy as conceived by the Western powers. Fascism and communism were claimed by their followers to be utterly opposed to each other. Germany had claimed to be Japan's friend and Russia's enemy, yet here she was making an alliance with the enemy and turning

Snow Festival at Sapporo City, Hokkaido. Sapporo is the site of the 1972 Winter Olympic Games.

Meiji Shrine, Tokyo, built in pure Shinto style. The shrine is dedicated to the Emperor Meiji and his consort and was built in 1920.

her back on the friend. Germany and Japan's joint stand against communism had been formulated in 1936 in the Anti-Comintern Pact, which was not to expire until 1941. There was a secret clause agreeing on armed co-operation if either power came into unprovoked conflict with the Soviet Union.

In 1939 Japan saw herself faced by foes on all sides and without an ally. America had been pressing for an end to aggression in China; instead Japan occupied the Hainan Islands and the Spratley Islands. The British concession at Tientsin was threatened. The American reaction was to notify Japan that America would not be renewing the 1911 Trade Treaty, implying that economic sanctions might be under discussion.

Opinion in Washington on the possible effects of economic sanctions was divided, but business houses were requested by the government not to export to Japan eleven items which were useful in time of war, and which America wished to stockpile in anticipation of that event.

In Japan some were fearful of provoking America further, but many were convinced that a short, sharp conflict could only end in Japan's favour. Ribbentrop in Berlin argued that Germany's pact with Russia would enable her to fulfil her aims in Europe without delay, then she could turn to help Japan to become master of the whole Pacific area.

The Reasons for Japan's Aggressive Policy

1. Lack of space for Japan's large population was probably the root cause, giving rise to a general feeling of unrest in the country. Immigration was severely restricted to North America and impossible to Australia. Nearby China had large unpopulated areas in Manchuria. It was easy to persuade people that here was a natural field for Japanese expansion.
2. Raw materials are deficient in Japan and plentiful in neighbouring countries.
3. There was a distrust of Russia's ambitions and intentions towards north China and Korea.
4. The army was self-confident and anxious to make use of its power.
5. The democratic government failed completely to curb army elements and it was helpless in the face of military pressures.

6. There had been a growing disillusionment in Japan with the policy of the European powers. After the First World War there seemed to have been a succession of slights to Japan. American opposition to naval expansion was followed by Britain's unwillingness to renew the treaty with her former ally. The two English-speaking nations then appeared to range themselves together against Japan in support of China, without any attempt to consider what Japan felt to be just and reasonable claims. Japanese diplomats walking out of the League of Nations Assembly gave dramatic expression to their feelings.

7. The death in 1940 of the last of the *Genro,* Saionji, was the end of the old aristocratic group which had kept a restraining hand on events.

8. The apparent success of Hitler against the Allies seemed to confirm the optimistic view held by some of the army officers about their own likely progress against a similar foe.

20

The War in the Pacific

By 1940 Hitler had overrun half of Europe, and Britain seemed to be in no position to oppose Japan's demands. In September 1940 Britain agreed to close the Burma road at Japan's insistence. France had been forced to admit Japanese troops into French Indo-China (Vietnam) to enable them to stop supplies reaching Chiang Kai-shek.

In September 1940 Japan, Germany and Italy concluded the Tripartite Pact. The signatories recognised each other's rights to leadership in Europe and Asia respectively. They pledged themselves to co-operate in getting the rest of the world to recognise it too. They further undertook to assist each other by every means in the event of one of them being attacked by 'a power not involved in the European war or in the Sino-Japanese conflict'.

In March 1941 the Foreign Minister, Matsuoka, visited Rome and Berlin. On his way home he passed through Russia and made a neutrality pact with Stalin in April. The Germans had not revealed that their next victim was to be Russia. They were in fact trying to persuade the Japanese to attack Singapore to draw British attention to the East. The Foreign Minister was under the impression that Germany would then launch an attack on England. Actually, Matsuoka was scarcely back in Japan reporting his success, before German armies were on their way to Moscow in June.

Hitler now urged his ally to attack Russia and repudiate the recently completed agreement. Japanese opinion was divided as to the best course of action. One deciding factor as far as the army was concerned was the severe mauling it had received at the hands of the Russians at Nomon-Han in August 1939.

Instead of assisting Japan, the Germans had chosen this time to announce the Molotov-Ribbentrop Pact and ally themselves with Russia for an attack on Poland (September 1939). Obviously European allies only considered their own interests, so Japan should do the same, making the best of the situation as it existed.

Some Japanese felt that Hitler's suggestion was sound—a quick victory might be obtained over the Russians while they were fighting for their lives on the Western front. The alternative was a thrust southward. Here less opposition was anticipated because the colonial powers were already involved in Europe. A major consideration was that the raw materials which were vitally necessary to Japan's successful prosecution of the war could be found in South East Asia.

Japan had already formulated in 1940 a statement of her conception of her role in Asia. A radio broadcast had stated that: 'The countries of East Asia and the region of the South Seas are geographically, historically, racially and economically very closely related to each other ... the destiny of these regions is a matter of grave concern to Japan in view of her mission and responsibility as a stabilising force in East Asia.'

Japan saw herself as the leader of an 'East Asian Co-prosperity Sphere'. It was clear that, as Hitler overran Europe, the colonial powers would no longer be in a position to protect their dependencies in Asia. Whether Germany proved a reliable ally or not, Japan could profit by German victories. On the whole, the thrust southwards seemed to be the best proposition.

The first to feel the effect of the decision were the French in Indo-China and the Dutch in Indonesia (the Dutch East Indies). The French were forced to allow Japanese troops into their Asian territory on the pretence that they were there to prevent supplies getting through to Chiang Kai-shek. The confusion caused by the fall of France and Holland before the Germans gave the Japanese a further advantage.

The Dutch were faced with demands for oil supplies, but conceded only a part of what was asked for at the request of the United States. A plentiful flow of oil from Indonesian sources would have nullified the American embargo on oil exports to Japan.

The main object of the Japanese army at this stage was still the conquest of China, where Japanese troops were continually in battle. Without oil the army and navy would soon be crippled. Admiral Nagano, Chief of Naval Staff, informed the Emperor that Japan could only last eighteen months without an alternative oil supply. Prince Konoye was therefore anxious to make an agreement with the United States and the Japanese Ambassador in America was instructed to try to improve relationships. The Ambassador was still under the impression that his negotiations for a peaceful settlement with America reflected the intentions of the Japanese government, when the new Cabinet under General Tojo was putting finishing touches to the plans for a surprise attack on the American base at Pearl Harbour.

On 8 December (in Hawaii it was Sunday, 7 December) 1941, waves of bombers from aircraft carriers began to attack the American Fleet in Pearl Harbour. At the same time American planes in the Philippines were destroyed before they could take off. Siam was invaded from Indo-China. Troops landed at Kota Bharu in Malaya, and Singapore, Hong Kong, Borneo, Guam and Wake Island were suffering their first air raids.

The British and Americans were taken completely by surprise and even the Japanese themselves were amazed at the speed of their advance. In ten weeks the Japanese armies travelled down the excellent main roads built in Malaya by the British. The guns of Singapore faced southwards ready to oppose an attack from the sea. No one had ever dreamed of an enemy thrust from the north over the causeway. All was confusion and British reinforcements had barely landed before they were captured.

In Australia a plan for evacuating the continent north of Brisbane and holding a line of defence there was seriously discussed.

From Malaya the Japanese quickly overcame the Dutch resistance in Indonesia. Their next objective was Papua and New Guinea. Here their career was brought to a halt. In May 1942 a Japanese squadron escorting troopships round east New Guinea for an attack on Port Moresby was engaged by an American force. This engagement, known as the Battle of the Coral Sea, is of vital importance. The Japanese were so mauled that they gave up all hope of attacking Port Moresby from the sea, and formidable mountain ranges cut off the town from an attack by land.

In June another naval battle, the Battle of Midway, regained the initiative in the Pacific for the United States. Though first losses had been staggering, the United States showed a determination to rebuild her fleets and fight back. Fortunately there were no aircraft carriers at Pearl Harbour at the time of the surprise attack, so that they and their cargo of planes were available for immediate retaliation.

The Australian forces in New Guinea stemmed the Japanese advance from Dutch New Guinea (now West Irian) over the Owen Stanley Ranges and eventually pushed the Japanese forces back to the coast.

The Americans began a systematic 'mopping up' of Japanese-held islands and pushed nearer and nearer to the Japanese homeland. The Japanese now found themselves with many isolated garrisons to be supplied and maintained. Landings by both sides were accompanied by heavy losses. While the Japanese dared not leave any bases unfortified, the Americans began to attack only the main ones, leaving the others to be starved out by sea blockade now that the Japanese navy had been reduced to a fraction of its former size.

The Japanese claimed to be supporting Asians against colonialism, and certainly the 'colonials' were rapidly captured and their regimes ended, never to be fully revived. The French and British in Indo-China and Malaya and the Dutch in the Dutch East Indies were defeated so quickly and completely that their prestige could never be restored. Some of the 'freed' Asiatic people were inclined to look on the Japanese as liberators, but the majority found them far more cruel and oppressive than any colonial government of Westerners. The conquerors were also surprised by the amount of loyalty that many Asians gave their former masters. Some may have remembered past insults, but many recalled past kindness and contrasted it with the Japanese attitude. To the large Chinese groups in the captured territories the Japanese were from the start nothing but ravishers of their homeland, China.

An Indian Independence League was set up with Japanese

encouragement. Plans were drawn up for recruiting an Indian army to fight alongside the Japanese. A provisional government of 'free' India was set up under Subhas Chandra Bose. While some Indian soldiers fought bravely on the British side, others felt more sympathy for the invaders.

By 1944 the Japanese could see the Allied net drawing around them. The Americans were bombing Japanese cities, flying over to China, refuelling there and flying back again. In 1944 a determined thrust by the Japanese into south China captured some of the airfields. An even more ambitious plan was to cut off India from China, so that supplies could no longer be flown in via Assam. Subhas Chandra Bose assured his allies that India would break out in revolt against the British as soon as there was a chance. In the Battle of Kohima-Imphal this scheme was foiled. The Japanese were soundly defeated. The attitude of the Indians was never put to the test.

In June 1944 the Americans invaded the Marianas and went on to Saipan, only 1,300 miles from Tokyo.

In October the Japanese were defeated at the Battle of Leyte Gulf. An attack was launched on the Philippines and bombers from the Marianas were able to reach Tokyo. As if the gods too were against the Japanese, a severe earthquake in Nagoya damaged industries engaged in war production.

The End of the War

In 1945 some of the bitterest fighting of the war took place as the Japanese prepared to defend their homeland. They resisted strongly in the Philippines, and the eight square mile island of Iwojima, 900 miles from Tokyo, was defended to the last man by 23,000 Japanese. The capture of Okinawa in the Ryukyus cost America 39,000 casualties.

The Japanese were now resorting to the use of *Kamikaze* (Divine Wind) pilots, who crashed their load of bombs into ships without a hope of saving themselves. The shortage of planes was nothing compared to the shortage of trained pilots and the Kamikaze were only boys who knew nothing of flying except how to take off.

Germany and Italy were both defeated by this time and the Allies could turn their undivided attention to the one remaining enemy. The Soviet Union announced that it would not renew its neutrality pact with Japan when it expired in April 1946, a pact which up to this time it had suited both signatories to observe.

Although the Japanese did not know it, Stalin had promised at the Yalta Conference that he would attack Japan when Germany was defeated.

Desperately the Japanese offered South Sakhalin, the North Kuriles and North Manchuria to Russia in return for her neutrality and oil supplies. The oil was again in desperately short supply because the Allies had retaken the South East Asian oilfields. The Japanese also asked Russia to become their mediator for peace terms.

July 1945 the American President, Roosevelt, and the British Prime Minister, Churchill, issued a proclamation offering Japan 'prompt and utter destruction' unless they surrendered. Japanese cities and shipping were under constant, ruinous attack by now from the air, but the Japanese armies in Manchuria, Siam, Malaya, the East Indies and Indo-China were still formidable.

On 6 August 1945 an atomic bomb was dropped on Hiroshima. On 8 August Russia declared war on Japan and on the following day another bomb was dropped on Nagasaki.

The Cabinet, under Prime Minister Suzuki, argued whether to sue for peace, and even after the two atomic bombs some wanted to fight on. Finally the Emperor was asked his opinion. He stated without hesitation that Japan must surrender. The members of the Cabinet, however, were afraid that the Allies would insist on the Emperor being subject to the orders of the Allied commander, even that they might remove the Emperor from his position. The more realistic advisers felt that any terms offered by England and the United States would be better than those of Russia, who was advancing from the north.

On 14 August the Cabinet decided to agree with the Emperor and accept the Allies' terms. The following day the Emperor was to take the unprecedented step of a broadcast to the nation announcing surrender. Early that morning a group of army officers invaded the palace to search for the Emperor's speech. Fortunately the general in command at headquarters in Tokyo was able to persuade them to give up the attempt. The Emperor was able to deliver his message to a stunned people, and the war was over.

Reasons for Japan's Initial Success in the War
1. The Japanese were prepared and took the Americans by surprise. Although the situation had been deteriorating for some time and American intelligence had been warning their government of the possibility of a surprise attack on Pearl Harbour, no precautions were taken. In January 1941 Rear-Admiral (later Vice-Admiral) Bellinger wrote to the Chief of Naval Operations criticising the lack of preparedness in Hawaii, 'an important naval advance outpost'. In March 1941 he and General Martin, commander of the Army Air Forces in Hawaii submitted an analysis of probable Japanese strategy based on a study of Japanese history. They stated: 'In the past Japan has never preceded hostile action by a declaration of war . . . Japanese submarines and a fast raiding force may arrive in Hawaiian waters with no prior warning from the US Intelligence Service.'

Admiral Yamamoto, Commander of the Japanese navy, from a similar study of naval history, had reached the same conclusion —that a surprise attack dealing a crippling blow at the American Pacific forces was essential for the success of the plans of the Japanese army. Actually he envisaged a declaration of war preceding the attack. The diplomatic document which would have warned the Americans that war was imminent was delayed in delivery because of the difficulties of translation and the time taken up on such purely secretarial details.

Martin and Bellinger had pointed out that the Axis powers frequently took advantage of holiday periods, such as weekends, to launch an attack, yet Pearl Harbour was caught enjoying its usual Sunday relaxation.

Even with the element of surprise the American anti-aircraft guns on the ships in Pearl Harbour were quickly in action. A more acute awareness of the situation politically might have saved the day. As it was, no American aircraft carrier was caught in the attack. Although America lost 230 planes on the airfield, the damage to shipping and installations turned out to be comparatively minor.

The effect on the American public, however, was electric. Every effort was now turned to prosecution of the war. The advocates of non-intervention disappeared overnight.

2. The Japanese army was well prepared for war and had seasoned fighters who had practised their profession in China. The defence forces in Malaya, on the other hand, were softened by long years of peace.

3. The fanatical bravery of the Japanese soldiers was due to the fact that it was considered an honour to die for the Emperor and a dishonour to be taken prisoner. The apparent willingness of the 'white' troops to surrender was incomprehensible to the Japanese and added to their self-confidence.

4. There was a certain amount of sympathy towards the Japanese at first from the indigenous populations of the invaded countries. Siam, for instance, was co-operative.

5. The defeat of the French in Europe left French Indo-China an easy prey.

6. The Western powers were fully engaged in warfare in Europe. When reinforcements were sent, it took them so long to arrive that the Japanese were in control of Malaya before their journey ended. Neither France nor Holland was in a position to maintain troops in Asia in adequate numbers for the new crisis.

Reasons for Defeat
Japan had to stake all on quick success because her oil supplies were dangerously low and other raw materials deficient. As Japanese armies moved further and further from the homeland, keeping them in arms and food was a major problem. When the United States recovered from the shock of Pearl Harbour, she was able to recoup her losses and soon had superiority at sea and in the air. Japan's vulnerability was then evident.

21

Foreign Occupation

No one was quite sure how the Japanese people would react to the American forces, or vice-versa. Japanese troops guarded the route taken by the victorious general, MacArthur; their backs were turned to the road and their guns were ready in case anyone launched an attack.

Suicides were common. Small groups of fanatics continued to defy the Emperor's orders to lay down their arms, but were gradually brought to obedience. Princes of the royal house were sent to outlying Japanese armies to tell them what had taken place and persuade them to give up fighting. Some isolated units continued, long after the cessation of hostilities, to regard the American troops' broadcasts to them that the war was over as mere propaganda.

In Indo-China and Indonesia, Japanese troops had to be rearmed and asked to keep order among the local inhabitants when anarchy threatened to disrupt the countries.

In September 1945 the Instrument of Surrender was signed on the United States warship *Missouri* in Tokyo Bay.

General MacArthur became Supreme Commander of the Allied Powers. His administration is known as SCAP, from the initials. A new Japanese government was formed under Prince Higashikuni, the Emperor's cousin. The Foreign Minister, Shigemitsu, persuaded MacArthur that the orders of SCAP would be more effectively carried out through the existing administrative agencies than through officers of a foreign military power.

There were many among the Allies who felt that the Emperor and the whole political structure of Japan should be swept away. A study of the Meiji Constitution seemed to show that the Cabinet was but an instrument of the Emperor's will and the

House of Representatives merely a confirming body. In fact, as we have seen, this was not the case. A strong-willed head of state might have kept control, but a mild-mannered sovereign like Hirohito, who preferred the study of marine life to politics, could not combat the machinations of the military clique. With MacArthur supervising, the Japanese government could work quite efficiently. The General realised the unifying power of the imperial throne, especially when communist agitation began to sweep the country.

Russia clamoured for the removal of the Emperor, in fact advocating that he should be shot as a war criminal. Fortunately for Japan, Russia's late entry into the war meant that Russian troops were not in occupation as they were in a substantial part of Germany.

Some of Russia's demands had to be met. Political prisoners, many of them communists, were released from prison. The Communist Party was legalised and a new Labour Union Law, fostering the growth of trade unions, which had been illegal during the war, was passed. Russia proposed that the industrial estates of Zaibatsu should be broken up. Titles, except those of the royal family, were to be abolished along with the House of Peers, which was thought to be a stronghold of reaction. The Home Ministry, which had conducted the ultra-nationalist state propaganda through its control of schools, police, and religion, was to be abolished.

At the Moscow Conference in December 1945, Russia had tried to insist that Marshal Vassilievsky should be joint commander with MacArthur. She also wanted control of Hokkaido. MacArthur firmly rejected both proposals. From the beginning he regarded his former allies with suspicion and was always sympathetic towards the persons and institutions they attacked.

In 1946 the Japanese government under Baron Shidehara came to an agreement with SCAP on a new constitution. Before the discussions on the various clauses had been terminated, a general election was held and Yoshida Shigeru became the new Prime Minister. He had been imprisoned in 1945 for advocating an end to the war, so politically he was acceptable to the occupying powers. Thus Yoshida was the man who was to guide the new constitution through its initial difficulties.

The Japanese Constitution of 1946
This constitution is prefaced by a statement of principles strongly reminiscent of the preamble to the constitution of the United States, which, of course, served as a model.
1. Instead of a President, the Emperor stands at the head of the state. He is termed 'the symbol of the state and of the unity of the people, *deriving his position from the will of the people*, with whom resides sovereign power'. The document itself begins: 'We, the Japanese people', instead of being graciously presented to the people by favour of the Emperor.
2. Ministers were to be responsible to parliament. The Prime Minister is chosen by parliament and he appoints the other ministers, a majority of whom must hold seats in parliament. The House of Peers was abolished and replaced by a House of Councillors, the members of which are elected for a term of six years, half going to the polls every three years.
3. All the ministers must be civilians.
4. Votes were given to women for the first time and the voting age for men was lowered to twenty.

Perhaps the most curious feature of the new constitution is Article 9, in which the Japanese people renounce forever warfare and 'the threat or use of force as a means of settling international disputes . . . land, sea and air forces, as well as other war potential, will never be maintained'.

Administration under SCAP
General MacArthur was provided with an advisory council in Tokyo of representatives from the United States, the Soviet Union, China and the British Commonwealth (England, Australia, Canada, New Zealand and India).

In Washington was the Far Eastern Commission with representatives of the United States, Britain, the USSR, China, France, the Netherlands, Canada, Australia, New Zealand, India and the Philippines. Its instructions went to MacArthur via the United States government.

The 1946-48 War Crimes Trials
Prominent statesmen, business men and military leaders were tried for crimes against peace, conventional war crimes and

crimes against humanity. Seven were condemned to death in Tokyo including the ex-Prime Ministers Tojo and Hirota. Prince Konoye, who was Prime Minister at the end of the war, committed suicide. Eighteen more men were given prison sentences. Approximately nine hundred executions were carried out in various parts of South East Asia, of persons found guilty of cruelty to prisoners of war or civilians.

All officials who had held office during the war in Japan were supposed to be removed, but many were quietly returned to their desks, because it would have been impossible to carry out such a sweeping change without chaos. The dissolution of the Zaibatsu would have been fatal to a country already hard hit by war. Despite Russian hostility, they were allowed to remain.

Educational Reform
The Americans felt that the old school system had been used for teaching militaristic and ultra-nationalist ideas. All this was to be done away with and a system similar to that of the United States was introduced.

At first everything was chaos. There were not enough teachers to man the new schools and universities; there was a breakdown of traditional Japanese respect for older people; and many of the new teachers were communist sympathisers, who took the opportunity of indoctrinating their pupils. Because of haste to get rid of the rigid discipline associated with the old days, the educational reforms produced a decline in standards both of behaviour and of scholastic achievement. The Japanese were urged to drop their cumbrous system of writing and adopt the Western alphabet. At first all these suggestions were carried out, but the romanisation of the writing has been quietly dropped and other features have been remodelled to suit the Japanese temperament.

Reform of the Police
The centralised police force was disbanded and small independent units constituted, responsible to local assemblies or prefectural governors. America was once more the model. In 1954 the country reverted to a centralised police force. Appoint-

ment of the chief of police and management of the system is entrusted to the Public Safety Commission.

Agricultural Reform
Plans for agricultural reform were drawn up by the Japanese Government and by the staff of GHQ. The one finally put into practice was substantially the work of W. MacMahon Ball, the British Commonwealth representative, who since 1949 has been Professor of Political Science at the University of Melbourne and is an acknowledged expert on Asian affairs.

As a result of the land reform, production was increased and the situation in the countryside was improved both morally and financially. Although the Japanese reverted to many of their old methods of organisation when the Army of Occupation was removed, the reform of agriculture worked so well that it has been maintained.

The terms of the agricultural reform were:
1. All land belonging to absentee landowners was to become the property of tenants.
2. Individual tenant or private holdings were to be limited in size.
3. Any land beyond the allowed limit was to be sold to the government, and was resold at reasonable prices to peasants.
4. Landowners were compensated by government bonds.

22
Japan After the War

The legalisation of the Communist Party and the freeing of political prisoners had immediate effect. In the disturbed state of the Japanese economy and moral life at the end of the war, the communists found a ready field for their propaganda. In October 1945 communist-inspired demonstrators occupied the premises of the largest Tokyo newspaper, claiming they wanted to 'democratise' the management.

In February of the following year the executives of a coal-mine in Hokkaido were given a 'peoples' trial' by miners. Strikes, demonstrations and terrorist acts were frequent. Nozaka Sanzo, a Japanese communist who had been exiled for years, was welcomed back in a public ovation by his enthusiastic supporters. Communists quickly gained domination of the new trade unions and encouraged strikes.

By 1947 GHQ had decided to intervene. A general strike was forbidden by MacArthur. Later a law was passed forbidding strikes by government employees, such as postal workers, and substituting a National Personnel Authority to arbitrate on wages and other causes of dispute.

In 1952 the Prime Minister, Yoshida, was given the power to halt labour disputes and refer them to a Labour Relations Commission. The bill was passed against noisy opposition in parliament from the left-wing groups. In the following year a bill to prevent the disruption of industry by strikes of coal and electricity workers survived a stormy passage. Strikes are still common, however, and they have been rationalised in a peculiarly Japanese way. It has become the custom for a wave of strikes to hit the country each spring, summer and winter to demand higher wages and higher mid-year and winter bonuses.

Wedding of Crown Prince Akihito and Crown Princess Michiko. They were married in April 1959 in a traditional ceremony.

The Imperial Family (left to right): Crown Prince Akihito; his younger son, Prince Aya; Prince Hitachi; the Empress Nagako; The Emperor Hirohito; Prince Akihito's elder son, Prince Hiro; Prince Hitachi's wife, Princess Hanako.

New Year's Day begins with a toast, which shows that modern Japanese are still concerned with tradition.

Family watching television—a typical evening scene in modern Japan.

The Japanese say: 'It is strike season.'

In parliament the first elections returned thirty-five Communist Party Members. By 1951 not one of these seats was retained. In 1967 there were once more five Communist members in a total of 486 representatives. The Prime Minister, Yoshida Shigeru, (1948-54) had considered the advisability of outlawing the party, but felt that Japanese common-sense was obviously reasserting itself. Many intellectuals, especially writers, joined the communist outcry against the continued presence of the Army of Occupation in 1950 and against the Korean war which broke out that year. The government removed some 22,000 suspected communist sympathisers from positions in press, radio and industry, because it was discovered that there was a plan to use these people to create so much strife in Japan that the American forces would not be able to be withdrawn for transfer to Korea. They would have to remain to keep order.

The San Francisco Peace Treaty

In 1951 the San Francisco Peace Treaty made Japan once more a free and independent nation, recognised as such by all but the USSR.

General MacArthur had always been against a prolonged occupation. He felt that it could only lead to bad relationships between the subject people and the controlling power. The outbreak of the Korean war meant that America could no longer afford to keep so many men in Japan. The Americans also felt that they would appreciate Japan as an ally, but the constitution specifically rejects warfare for the Japanese. The interpretation was so far relaxed to allow Japan to form a national safety force from the national police reserve. It was ostensibly for self-defence only, after the American troops were withdrawn.

The Peace Treaty, apart from arranging the withdrawal of the occupation forces, specified that the Japanese should negotiate agreements on reparations for war damage with the countries which had suffered her depredations during the war. It was agreed that America should retain control over the Ryukyus and the Bonin Islands.

Russia refused to discuss any return of the islands her forces had taken over before the cessation of hostilities.

The US-Japan Security Treaty recognised the United States' bases in Japan and promised armaments from America for Japan's new security forces.

Russia, Australia and New Zealand expressed fears that the small army, navy and air force that Japan was to have would form the nucleus for a larger, dangerous accumulation of men and weapons. The latter were reassured by the formation of the Anzus Pact for mutual defence. The name comes from the initials of the signatories: Australia, New Zealand and the United States.

The Current Situation in Japan

From 1952 onwards Japan has overcome some of the political unrest from which she suffered but there are still student risings and communist demonstrations. The people are easily aroused by any threat of atomic contamination. The American base at Okinawa has been a continuing sore point, though Okinawa is now to be returned to the Japanese in 1972.

Many Japanese, whether communist or not, feel that China is Japan's natural partner. Alliance with America makes friendship with China difficult. On the other hand, Chinese atom bomb experiments, which contaminate the atmosphere around Japan, are both a danger and a warning to that country. Her position between China, Russia and America is unenviable. She can only hope that this time her statesmen are backing the winning side.

In economic affairs Japan is back among the top nations again. The Zaibatsu are as powerful as ever. With the backing of the government, which we have seen is one of the distinctive features of the country's industrial and commercial organisation, the business combines have placed their country second only behind the giants of America and Russia.

Japanese businessmen have recognised the danger of being too closely tied to the American market. In the 1960's a new trend became clearly visible. Australia is rapidly changing her former close trading partnership with Britain for one with Japan. The two small island countries are very similar in products and needs, but Japan is much nearer to Australia than is Britain. It has a large industrial population requiring the raw materials which Australia produces, and which continue to be discovered as the vast continent is explored for minerals.

The Japanese people take increasingly large quantities of Australian primary produce as their tastes in food change. Japan is already Australia's most important customer for wool, while Japanese capital and heavy equipment is employed in developing Australia's natural resources. Both countries look forward to further markets in South East Asia opening up to them. No longer are Japanese products merely cheap imitations of Western goods; strict government supervision over exports ensures that quality is maintained.

As Japanese industry has learned from its mistakes, members of parliament must also learn that brawls in the Diet and public scandals about corruption are not conducive to the growth of democratic rule.

Political assassinations are still far too common. In 1960 the leader of the Socialist party, Asanuma, was assassinated. Prime Minister Kishi had to face demonstrations and student riots that prevented the visit of President Eisenhower to Japan in June 1960, the occasion being the ratification of the Japanese-American Security Treaty. In the following month Kishi was the subject of attack by a would-be assassin. Prime Ministers Yoshida, Hatoyama and Ikeda all survived attempts on their lives. Public office in Japan is not for the faint-hearted.

The Emperor has willingly stepped down from his former position as a deity. Young Japanese would find the idea ludicrous. His son, Crown Prince Akihito, was brought up on democratic lines, and his education was supervised by an American Quaker, Mrs Vining. In 1959 he showed his independence and his affinity with the new Japan by choosing as his own wife a girl of no very high social pretensions, whom he had met on the tennis courts.

Japan's prosperity is due to the industry of her people. Alone among Asians the Japanese have voluntarily restricted their families to a size dictated by their incomes. The population is still too large for the country but the situation is not hopelessly chaotic.

Japan's education system is a model now that the post-war difficulties are being overcome. Schools are provided for all by the state. Parents are expected to pay for extras: books or music lessons.

Graduates are turned out from the universities in thousands and are absorbed into industry and commerce. Possession of a degree is a necessary qualification for a growing number of positions. The women too are taking advantage of their new freedom and are contributing to the development of the country.

The prosperity of Japan seems assured (1968), but four questions remain to be answered in the future:

1. Can Japan manage to steer a course between the rival communist and Western camps without hopelessly antagonising one or the other?
2. Can Japan cope with her increasing population without a growth in territory, or without a relaxation of the immigration laws of such empty countries as Australia? This to the Japanese is a question which must be answered very soon.
3. Will militarism raise its head again when economic or population pressures make themselves felt seriously?
4. What part will be played in the future by the growing Buddhist political group known as Soka Gakkai? Its aims are at present obscure, but it recruits vigorously and unscrupulously, and specialises in mass meetings and an adulation of the leader that is strongly reminiscent of the birth of the Fascist movement in Europe.

23
Conclusion

Japan today still has the same problems that faced her at the beginning of the century: a teeming population and a lack of raw materials. The solution of the militarists was to conquer more territory. However, they met with only temporary success. Besides defeat in war, they met with lack of co-operation from some sections of their own countrymen and from their neighbours.

Manchuria was seen not only as a source of raw materials, but also as a place where Japan's surplus population could find a home. Unfortunately for the scheme, the Japanese as a race are reluctant to emigrate; even Hokkaido holds little attraction for them. The amount of emigration was never enough to alleviate the population problem. The average Japanese prefers to stay at home and limit his family to a size he can support and educate easily.

The southern thrust from Japan during the war was mainly to acquire oil and other raw materials. The Japanese saw themselves as leaders in what they called 'the Great East Asia Co-prosperity Sphere'. However, the emerging nations of Asia were not prepared to exchange one master for another and the cruelty of the Japanese soldiers soon disillusioned the 'newly-liberated' countries.

The militaristic approach has been discredited. The economic approach seems much more hopeful. The need for raw materials is being met by peaceful trade. Shrewd business deals have replaced the brilliant tactics of the generals as a means of advancing the country.

The population problem remains. It is a problem shared by the rest of the world. Here again the Japanese approach is realistic—voluntary family planning. The solution is not a new one in Japan and is not repugnant to the people on moral or religious grounds as in other countries of the East.

The emphasis on education also has an effect on population. It keeps young people at school and at university so that the marriage age is much higher than formerly. The highly educated work force gives Japan an edge over her competitors in trade and must be counted as an important factor in the post-war success of the country.

Another factor in Japan's favour is the conservation of natural features which prevent soil erosion. Manchuria and Korea are both today benefiting from the Japanese occupation because the reafforestation programmes in those countries are beginning to come to maturity. The same careful conservation programme assists Japan to make the most of the soil resources she has, scanty though these may be. This must be contrasted with the puny efforts in this field of such countries as Indonesia and the Philippines.

Perhaps the greatest factor in Japan's resurgence is the great sense of nationality that must be ascribed in part at least to the seclusion period in her history. Turned in on themselves for so long, the Japanese have a sense of belonging to one people, one family almost, that can scarcely be appreciated by the other polyglot races that people the earth.

In contrast to this, the younger generation seems to be against the old standards of behaviour. The policy at the end of the war was to discourage the study of history, which many felt had been the basis for the power of the militarists. It certainly was used by them to foster belligerent nationalism, but the consequent divisions created between young people and their elders has been unfortunate. Student unrest is only the most visible of these divisions. It is all the more disturbing because it appears to be so out of character, so different from the Confucian basis of society. However, as a recent television commentator noted, the unruly student seems to become quiet, conservative and hardworking as soon as he becomes an adult and is reabsorbed into the traditional stream of Japanese life.

Japan's geographic position, as we have seen, is fraught with peril, but modern methods of transport have removed much of the significance of mere proximity to a possible attacker. The big powers can now send their missiles to any part of the globe.

It is probably more significant in the world picture that capitalism, or the Western way of life—call it what you will—should have a moral victory in Japan. Certainly here is an Asian country which is flourishing and holding a prominent position among the chief nations of the world without recourse to either of those panaceas of economic distress, socialism and communism.

Appendices

Appendix I: The Forty-Seven Ronin
No book on Japanese history could be considered complete without an account of the most famous of Japanese events, the story of *The Forty-Seven Ronin*.

The term *ronin* literally means 'wave men'. They were warriors who were unattached to a lord because of some misfortune which had befallen him. The defeat of a lord in battle, or a reduction of his rice-based wealth for any reason, led to numbers of his retainers being turned adrift to fend for themselves. Sometimes they were taken over by another lord, but often they became poverty-stricken wanderers. It was ronin from the confiscated fiefs of the nobles hostile to Iyeyasu who rallied around Hideyori in 1614. They held the fortress of Osaka for nearly twelve months. Iyeyasu was only able to capture it after filling in the moat.

The ronin were in a peculiar position because, while they might be penniless, they still maintained their status as samurai. In the *Legacy of Iyeyasu* their privileges are listed: 'The samurai are the masters of the four classes. Agriculturalists, artisans and merchants may not behave in a rude manner towards a samurai . . . a samurai is not to be interfered with in cutting down a fellow who has behaved towards him in a manner other than is expected.' Thus, brutality to other men was sanctioned and anyone who failed to show proper respect to a samurai could be cut down on the spot, as the unfortunate Richardson was cut down in 1863.

The conscript soldiers of the Meiji restoration felt themselves to be the successors of the samurai and some of the harshness of the Japanese army, both to its men and to prisoners, must be ascribed to the inheritance of the warrior tradition.

The story of the *Forty-Seven Ronin* is so popular as a play, and recently as a film, that the Americans considered banning it during the occupation. Although the action of the plot is set in 1703 it was obvious that the drama had some special significance to the audiences.

Briefly the plot is as follows: a daimyo named Asano is insulted by another, named Kira. The incident is not described in detail in the play, but its very triviality is a good illustration of the petty matters which filled the minds of men during the seclusion period. Asano was not sufficiently acquainted with court ways so he was obliged to rely on Kira, the court official, to guide him through the procedure when he appeared before the Shogun. In return a gift was necessary. Unfortunately, Asano's gift was considered insufficient by Kira, who retaliated by allowing Asano to make embarrassing mistakes in court etiquette.

Asano felt the shame so deeply that he drew his sword and attacked Kira without waiting until they had left the Shogun's castle at Yedo.

To unsheath a sword at court was punishable by death. Asano was ordered to commit suicide. As a result of his death his retainers became masterless ronin. They resolved to avenge their overlord. Under a leading retainer named Oishi, they plotted their future actions and waited for a suitable time to attack Kira, who, of course, realised his danger and was on his guard. Those ronin who were not wholeheartetdly bent on revenge gradually dropped out of the plot and at last only forty-seven were left.

The ronin prepared for their final act of revenge with almost religious fervour, scenting the inside of their helmets so that their heads would be fragrant if cut off by the enemy in the course of the fight they were anticipating. In black armour they crept through a snow-storm to the wall of Kira's house in Yedo and the attack began. Kira's men were surprised and were unable to defend their lord, whose head was cut off in triumph by the Forty-Seven. They took it and placed it before Asano's tomb.

The Shogun was in a quandary. Public opinion was all on the side of the ronin, who had followed Confucian ethics and avenged the death of their feudal superior. However, private revenge was not to be tolerated in a well-governed country. It was finally decided that everyone would be satisfied if the remaining ronin committed ceremonial suicide. This they did en masse. They were buried with all honour next to their master. Their tombs, even today decked with offerings from their admirers, can be seen in the grounds of the temple of Sengakuji in Tokyo.

Opinions as to the influence of the story differ. The occupying Americans eventually decided it was harmless. However, in the militarist period before the last war, army units had the story read to them on the anniversary of the ronins' death as an example of the loyalty they should copy.

The story certainly extols two ideas that the Japanese people seem to hold and which appear strange to outsiders. First, the assassination of Kira is regarded as praiseworthy because the motives of the assassins were unselfish. Political assassinations in Japan are often excused for the same reason. Secondly, the idea of mass suicide as carried out by the ronin was copied as late as 1945, when groups of people committed suicide in front of the Emperor's palace after the news of the Japanese surrender. The prevalence of suicide in Japan is always remarkable to Westerners. To a Christian it is a sin, but in Shinto and Buddhism the idea of giving up one's life voluntarily is quite acceptable. The suicide by burning which is publicly performed by Buddhist monks and nuns seems a peculiarly painful way of death to choose voluntarily. It is, however, directly sanctioned by Buddhist teaching. Artistic works of the Nara period often show incarnations of Buddha burning his physical body for the common good. The purifying property of fire was a frequent theme of Buddha's sermons.

In modern Japan young people kill themselves in love pacts or because they fear to fail an examination.

The film version of the story of *The Forty-Seven Ronin* is to Western eyes seriously overstocked with suicides, but to the Japanese the story has an unfailing appeal. The anniversary is still celebrated annually with film and stage performances to huge audiences.

There is a distinct resemblance between the actions of the

Forty-Seven and those of the followers of Yoshida Shoin and Hashimoto, who were executed for plotting against the Shogunate in 1858. These two were among the many Japanese who objected to any contact with the foreigners. On their deaths their followers swore to have their revenge. Ronin were responsible for many murders of foreigners and pro-foreign Japanese, culminating in the assassination of the Regent. Sir Rutherford Alcock, the British Ambassador, in 1863 blamed the wave of murder on the constant praise for the example set by the Forty-Seven Ronin.

Yoshida and Hashimoto's followers joined with the western clans of Satsuma, Choshu, Tosa and Hizen in bringing about the fall of the Shogun. The latter's supporters then saw themselves as heroic ronin and many of them found an honourable death for themselves on the battle field at Toba-Fushimi.

Saigo's rebellion in 1877 also employed ronin and destitute samurai. It is obvious that the idea of rebellion, sanctioned by misguiding feelings of loyalty, is a disrupting influence still in Japanese politics.

At the end of the nineteenth century the notorious 'Black Dragon Society' invoked the example of the ronin to excuse its extreme nationalism. The members were anti-foreign and advocated a policy of territorial expansion on the mainland. Their views not unnaturally coincided with the sentiments of the army which used them to gain information about Korea and China.

The ronin, ancient and modern, represent the feudal, decentralised aspect of Japanese life. The Emperor is the centralising influence. The idea of parliamentary government is strange to the Japanese and has no roots in their history. The 1946 constitution emphasised this by its insistence that the Emperor had no part in granting the new constitution, so the 'ronin element' can attack parliament with impunity. The danger is that loyalty to a cause may be invoked as a reason for any kind of violence.

Appendix II: Will Adams

The career of Will Adams, besides being so romantic that it has been the subject of several novels, is an interesting illustration of the contacts between Japan and the West before the seclusion period.

Born in Gillingham, Kent, in 1564, Will Adams was apprenticed to a shipbuilder. At the age of twenty-four he was the master of a small ship and was one of the many who transported supplies to the English fleet as it pursued the fleeing Spanish Armada. He evidently preferred life at sea to shipbuilding and even after his marriage, he continued to make voyages. A typical Elizabethan seaman, Adams was not content unless there was an element of adventure in his trips. In 1593-95 he sailed with a Dutch expedition to try to find the North-East Passage that some were sure could be found round the north of Russia.

In 1598 he made the momentous decision to go on a Dutch expedition to the East Indies. From this voyage he was destined never to return.

The route taken was across the Atlantic and through the Straits of Magellan. Of the five ships that set out only four reached the Straits; the fifth disappeared in the Atlantic. Of the remainder, one returned to Holland, another reached the Indies and was captured by the Portuguese, a third was captured by Spaniards. The *Liefde*, with Will Adams on board, waited for a month at the rendez-vous then continued across the Pacific. In 1600 the ship reached Japan with only twenty-four of its crew alive and six of those died soon after landing.

They arrived during the power struggle that followed the death of Hideyoshi. Iyeyasu, hearing that the Dutch ship carried among its cargo matchlocks, cannon balls, gunpowder and chain shot, sent for a leading member of the crew. Will Adams was chosen and he negotiated with Iyeyasu the transfer of weapons that were to help Iyeyasu to defeat his opponents. In return the weary travellers were given a house in north east Kyushu and allowed to come ashore to recuperate.

Spanish and Portuguese missionaries were already established round Nagasaki and news of the recent arrival filtered through to them. A visit confirmed their worst fears—these were heretics and potential trade rivals. They urged Iyeyasu to get rid of the interlopers at once. Instead Iyeyasu sent for Adams to put his point of view. Fortunately for the travellers, Adams had picked up some Portuguese in the course of his sea-going and could make himself understood. The future Shogun understood the position very well when Adams explained that his country, England, was not on good terms with Spain and Portugal. He also understood the art of playing off one power against another, as he showed by his own victorious career.

The *Liefde* was brought round to Osaka and Will Adams was reunited with his companions. There they were able to persuade the Japanese to return their belongings and let them go. Most of them disappeared from history, but Adams was retained by the Shogun to build him a European-style ship and to teach him mathematics. Their relationship was so successful that Adams was made a tenant-in-chief and given a large country estate at Hemi (Miura district) near Yokosuka in South-eastern Honshu. He married the daughter of a posting station manager on the Tokaido and had a son and daughter.

Adams was now a person of some consequence and the missionaries, after trying in vain to get rid of him again, decided to make the best of a bad job and co-operate with him in trading ventures.

Some of the Dutchmen who had accompanied Adams made their way back to Holland and reported on the possibilities of trade. In 1609 two Dutch ships arrived at Nagasaki, then moved round to Hirado. Adams accompanied a delegation from them to visit Iyeyasu and they were granted a trading post at Hirado and the right to trade. The letters which Adams had entrusted to his Dutch friends to deliver to England were lost. His English wife was left to consider herself a widow, while English traders were uninformed of the opportunities they were missing.

In 1608 Adams was sent to the Philippines by the Shogun to negotiate terms for Spanish trade with Japan and three years later the Spaniards established an ambassador in the country. His aim was to get rid of the Dutch, but Adams was able to convince the Shogun that the Spaniards aimed at capturing the world. First, he said, they send missionaries, who convert the people to Christianity, then they send troops who are sure of a welcome from the converts. It is difficult to guess exactly how much influence Adams had on the Japanese potentate, but it is certain that 1614 saw Iyeyasu turn against the missionaries, ordering them to leave Japan.

Meanwhile further letters from Adams had reached England and in 1613 an English ship arrived. With Adams' assistance a trading post was granted at Hirado. The expedition leader was inclined to treat Will Adams as a common seaman. The latter realised that if he returned to England he would no longer be regarded as a nobleman, the valued friend of a ruler, and his instructor in various fields. The permission to return to visit his homeland was never used. Instead Adams was taken into the employment of the East India Company and his knowledge of Japan was used in furthering English commercial ventures.

In 1616 Adam's old friend Iyeyasu died. His son Hidetada, who had held the title of Shogun for years but had no real power while his father was alive, proved to be an enemy of the Christians and not particularly friendly towards Adams. The Englishman waited for days at the court for a reply to his requests for continued trading benefits for his countrymen. Hidetada wished to make it quite clear that times had changed. Trade became more and more difficult though Adams, as a Japanese nobleman, could negotiate freely and sometimes help his fellow employees of the East India Company. He died in 1620, and the English trading factory only survived another three years without his assistance. Only a few more years were to pass before all foreigners, except the Dutch, were to be excluded from Japan.

Will Adams memorial can still be seen at Hemi. Almost forgotten now, he can still claim to be 'the first Englishman in Japan'.

Appendix III: Soka Gakkai
The movement known as *Soka Gakkai* is viewed with disquiet by some students of modern Japan. Its great success in recruiting new members is feared to be a prelude to a move for political dominance. Soka Gakkai claims to have over 15 per cent of the Japanese population as members and these presumably all vote for the election candidates sponsored by the *Komeito* (Clean Government Party) established by Soka Gakkai.

Based on the teachings of Nichiren Daishonin, the thirteenth century Buddhist reformer, Soka Gakkai was established in 1930 by Makiguchi Tsunesaburo. It soon came into conflict with the militaristic government of the time, and many of its adherents spent the war years of 1941-46 in prison. The founder himself was imprisoned as a result of his opposition to the attempt to substitute State Shino for Buddhism as the faith of the Japanese people; he died in prison in 1944.

When peace returned to a shattered Japan, the new religious movement gathered adherents rapidly, though there is nothing really radical in the message. Living a life of moral rectitude as taught by Nichiren is to bring his followers happiness and prosperity in a tranquil world.

The critics of Soka Gakkai, however, claim that there is something sinister about the phenomenal success of the movement. Members are failing in their duty if they do not make converts and it is said that the more zealous sometimes resort to economic blackmail to encourage their neighbours to join. This, of course, is denied by the society.

In view of Soka Gakkai's record during the Second World War, it seems hardly fair to accuse the group of being militaristic. Unfortunately, the mass meetings and rythmic singing and movement, which are a feature of the religious celebrations, are strongly reminiscent of former Fascist practices. Undoubtedly, with the wrong leadership such a movement could be very powerful and dangerous.

Glossary of Geographical Terms

Several variations of names and spellings are in common use in books and maps dealing with Asia. To prevent confusion the following should be carefully noted:

Amboina (Moluccas ... Indonesia): Now known as Ambon.
Bonin Islands: called by the Japanese Ogasawarajima (*-shima* or *-jima* means island). In the *Readers' Digest Atlas* they are labelled Kazan Retto (*retto* means a group of islands).
Borneo: known to modern Indonesians as Kalimantan.
Dairen, Dalien or Talien: anglicised forms of the Chinese name. Talienwan is also found (*-wan* means bay). Dalny is the name given to the town by the Russians.
Edo or Yedo: now called Tokyo.
Formosa: the Portuguese name for what is now known as Taiwan.
Harbin: now Pinkiang by the Chinese.
Hokkaido: the name usually given to the northern island of Japan, which is properly named Yezo. Hokkaido (Northern Sea Gate) originally denoted Yezo and the Kurile Islands.
Hupei: sometimes spelt Hupeh. This is not to be confused with Hopei (Hopeh), the province around Peking.
Indonesia: used to be known as the Dutch East Indies.
The Kanto: the plain around Tokyo, in some books referred to as the Kwanto.
Kurile Islands: sometimes Kuril Islands. The *Readers' Digest Atlas* uses the Russian Kurilskiye Ostrova.
Kwantung Peninsula or Luta Peninsula: the southern tip of the Liaotung Peninsula. Not to be confused with the Chinese province of Kwangtung around Canton.

Kyoto: The old name was Heian-kyo.
Loochoo, Luchu or Liuchiu Islands: now Ryukyu Islands.
Manchuria, Manchukuo, Manchoukuo: in Japanese, Manchukoku.
Mukden: now Shenyang.
Nippon: Japanese for Japan.
Port Arthur: sometimes Lushun.
Ryukyu Islands: or Loochoo, Luchu or Liuchiu Islands.
St Petersburg or Petrograd: now Leningrad.
Sakhalin or Karafuto.
Szechuan or Szechwan.
Taiwan or Formosa.
Tokyo: old name Edo or Yedo.

N.B. Aikokusha, Aikokuto: The Japanese political party can be called by either name, as *sha* and *to* both mean 'association'. Aikoku means 'love of country' (e.g. Manchukoku).

Remember that in Japan and China the surname of a person is given first, as in this book. When speaking to a foreigner, the Japanese frequently turns his name around, but in introductions the correct form is to bow and say one's surname.

Further Reading

Most of the following books include extensive bibliographies.

Early Period (General)
Fitzgerald, C. P., *A Concise History of East Asia*, Heinemann, Melbourne 1966
Kennedy, Malcolm, *A Short History of Japan* (soft back: Mentor Books, New York 1964; hard back: Weidenfeld and Nicholson, London 1964)
Latourette, Kenneth Scott, *Short History of the Far East*, Macmillan, New York 1964

Legends
Crown, A. W., *Folk Tales of the World Series: Japan*, E. J. Arnold and Son, Leeds 1963
McAlpine, Helen and W., *Japanese Tales and Legends*, Oxford University Press, Oxford 1958

Modern Period (General)
Beckmann, George M., *The Modernisation of China and Japan*, Harper International Student Reprint, New York 1964
Buss, Claude A., *Asia in the Modern World*, Macmillan, London 1964
Dening, Esler, *Japan*, Ernest Benn, London 1960
Michael, F. W. and Taylor, G. E., *The Far East in the Modern World*, Methuen, London 1964

Special Aspects
Allen, G. C., *Short Economic History of Modern Japan, 1867-1937*, Allen and Unwin, London 1951
Ball, W. McMahon, *Nationalism and Communism in East Asia*, Melbourne University Press, Melbourne 1952
Barr, Pat, *The Coming of the Barbarians*, Macmillan, London 1967
Bowers, Fabian, *Japanese Theatre*, Hermitage House, New York 1968
Dunn, C. J., and Yanada, S., *Teach Yourself Japanese*, English Universities Press, London 1958
Herbert, Jean, *Shinto: At the Fountain Head of Japan*, Allen and Unwin, London 1967
Jackson, W. D., *Russo-Chinese Borderlands: Zone of Peaceful Contact or Potential Conflict?* Van Nostrand, Princeton 1968

Books Recommended for Further Reading

Kahin, George McTurnan, *Major Governments in Asia,* Cornell University Press, New York 1965

MacIntyre, Donald, *Battle for the Pacific,* Angus and Robertson, Sydney 1966

Morris, Ivan (Ed.), *Modern Japanese Stories,* Eyre and Spottiswood, London 1961

Mosley, Leonard, *Hirohito, Emperor of Japan,* Weidenfeld and Nicholson, London 1966

Ogata, Sohaku, *Zen Buddhism for the West,* Rider, London 1959

Potter, John Deane, *Admiral of the Pacific,* Heinemann, London 1965

Reischauer, R. Karl, *Japan: Government and Politics,* Thomas Nelson, New York 1939

Roberts, Denis Russell, *Spotlight on Singapore,* Times Press, London 1965

Smith, Bradley, *Japanese History Through Art,* Weidenfeld and Nicholson, London 1964

Statler, Oliver, *Japanese Inn,* Secker and Warburg, London 1961

Walsh, Len, *Read Japanese Today,* Charles E. Tuttle Co., Rutland (Vermont) and Tokyo 1969

Watt, Mildred, *Japan: Land of Sun and Storm,* Cheshire, Melbourne 1967

Yoshida, Shigeru, *The Yoshida Memoirs,* Heinemann, London 1961

Questions and Suggestions for Further Work

Chapter 1
1. Make notes on Hokkaido, Honshu, Kyushu and Shikoku, summarising their main geographical features, chief towns and products.
2. Draw a map showing the physical build of Japan. Note the plains and add the ocean currents which affect the climate.
3. Summarise under nine headings the geographic and ethnic factors in Japan's progress.
4. Locate the island possessions of Japan on the map, noting their position carefully in relationship to Japan proper and the mainland.
5. Make sure you know the exact location of Manchuria and Korea in relation to Japan. Check the glossary of geographical terms for alternative names and spellings in this area.

Chapter 2
1. Begin a time chart under the headings: Japan . . . Asia . . . Europe.
2. Make brief notes on: the Ainu; Amaterasu; Jimmu Tenno; Emperor Suinin; Empress Jingo.
3. What did the Japanese learn from China in religion, culture and government? (Buddhism, Confucianism; language; Nara, Kyoto; Taika, Taiho Laws—see special chapters.)
4. How did the Fujiwaras rise to power? What gave them the opportunity? How did they consolidate their position?

Chapter 4
1. How were the feudal estates built up?
2. Explain the terms *samurai, daimyo, bushido, seppuku.*
3. What factors led to the fall of the Fujiwara from power?
4. Briefly describe the Gempei War.
5. What was the Shogunate and what led to its establishment?
6. List five effects of the Mongol Invasions.
7. Bring your time chart up to date.

Chapter 5
1. Make brief notes on: Go-Daigo; Hojo Takatoki; Ashikaga Takauji; Kogon.

Questions and Suggestions for Further Work 141

2. Who were the 'false Emperors'? What was the solution to the problem?
3. Give a brief account of the centres of government in Japan so far mentioned and explain how they came to be chosen.
4. Look at a book on Japanese art for pictures of the Muromachi era. Read about 'Noh' in the chapter on drama.
5. Look up the chapter on religion for the Nichiren sect of Buddhists. Use your libraries to find out more about Buddhism.
6. What were the reasons for the decline in the power of the Ashikaga Shoguns?

Chapter 7
1. Bring your time chart up to date. Read chapters eight and nine and make sure you understand how Japanese history at this point was involved in world history.
2. Summarise the careers of Nobunaga, Hideyoshi and Iyeyasu.

Chapter 8
1. What countries were interested in trade with the East in the sixteenth and seventeenth centuries? What traces of this contract remain today?
2. What contact did Europeans have with Japan in these centuries?
3. Why did Iyemitsu cut his country off from the outside world? What contact was still allowed?

Chapter 9
1. Summarise the economic and social conditions of the Seclusion Era.
2. What were: the Go-Sanke; the Fudai; the Tozama?
3. Enumerate the attempts of the Tokugawa to alleviate economic distress.
4. Look up again Russia's geographical relationship with Japan. Locate Sakhalin and Nagasaki. What was happening in Europe at the beginning of the nineteenth century? Put this in your time chart.
5. Find out more about the Opium Wars in China.
6. Read *The Coming of the Barbarians* by Pat Barr (see Further Reading).

Chapter 10
1. Enumerate the reasons for America's interest in Japan in the nineteenth century.
2. Make notes on Russia's involvement in the same area. Refer to *Russo-Chinese Borderlands* by W. H. Jackson (see Further Reading).
3. What were the features of the Treaty of Commerce and Navigation of 1858 which caused discontent among the Japanese? Were these clauses reasonable?

Chapter 11
1. Give five reasons for the fall of the Tokugawa Shogunate.
2. Revise by reading *Japanese Inn* by Oliver Statler (see Further Reading).
3. Summarise the ways by which control had been kept over the daimyo by the more powerful Shoguns.

142 Japan

4. What were Satsuma, Choshu, Tosa and Hizen? Mark their spheres of influence on your relief map.

Chapter 13
1. Give three reasons for the voluntary surrender of their privileged positions by the daimyo.
2. What were the causes of the rebellion of the samurai in 1877?
3. Do you agree that 'a democratic system of government presupposes an educated electorate'?
4. Outline the early attempts to set up a new form of government after the Meiji Restoration. (The Charter Oath, Council of State, prefectural and urban assemblies, political parties, Ito Hirobumi's report.)

Chapter 14
1. What do you consider were the good and bad features of the Meiji Constitution?
2. What was the position of the Emperor after 1890? (According to the constitution and as result of the rescript on education.)
3. How did the old class system show through the new form of government?
4. What were the features which played into the hands of the militarists? (Weakness of parties: dislike for the merchant class as represented by Mitsui, etc.; Yamagata; apparent success of military policy; desire for national prestige.)

Chapter 15
1. Check up on the geographic position of the Kuriles, the Bonin Islands and the Ryukyus. Revise the summary of their history in Chapter 1.
2. Summarise Japan's relations with Korea up to 1895.
3. Summarise the clauses and results of the Treaty of Shimonoseki, 1895.
4. 'Like all peace treaties, this seemed to hold the seeds of future trouble.' How far is this true of the Treaty of Portsmouth?
5. Make notes on the career of Ito.

Chapter 16
1. Bring the time chart up to date. Note on the map the positions of Hawaii and the Philippines.
2. Enumerate the reasons for the worsening relations between Japan and the United States up to the outbreak of the First World War.
3. Find out where Japanese settlers have established themselves in North and South America. Have they made satisfactory immigrants?

Chapter 17
1. Summarise what you have learned about Mitsui and Mitsubishi.
2. Enumerate the differences you have noticed in this chapter between the Japanese economic system and the Western system.

3. What do you understand by Zaibatsu? Why were they unpopular among some sections of the community? What did they do to try to improve their image?

Chapter 18
1. What were the three main streams of political thought that emerged in Japan after the First World War?
2. What were the 'Marxist ideas' that attracted Japanese youth in the 1920's? What do you think was their appeal?
3. Make additions to your time chart.
4. Read the account of the 1936 army plot in *Hirohito, Emperor of Japan*, by Leonard Mosley (see Further Reading).
5. Make notes on the growing power of the militarists. Why do you think they were able to reach such a strong position?

Chapter 19
1. Summarise Japan's policy during the period when the Western powers were involved in the 1914-18 war.
2. What questions were discussed at the Washington Conferences?
3. Who or what were: the Kwantung Army; Henry Pu-Yi; the Young Marshal?
4. What was: the Manchuria Incident, 1931: the Lytton Commission?
5. Make notes on the attitudes of the United States, Britain, Russia and Germany towards Japan from 1936-40.

Chapter 20
1. Fill in the time chart.
2. Read *Battle for the Pacific*, by Donald MacIntyre; *Admiral of the Pacific* by John Deane; *Spotlight on Singapore* by Dennis Russell Roberts (see Further Reading).
This chapter has been kept brief purposely so that a general picture of the situation can be obtained without confusing it with too much detail. Plenty of readable books tell the story from all sides.

Chapter 21
1. Discuss the problem of the future government of Japan as if you were members of the Allied command at the defeat of Japan in 1945.
2. Read *The Yoshida Memoirs* by Shigeru Yoshida (see Further Reading).
3. Compare the 1946 Japanese Constitution with the Meiji Constitution.
4. Read the preamble to the Constitution of the United States of America, and compare it with the Japanese Constitution.
5. Summarise the changes made in Japan immediately after the war. How successful were they?

Chapter 22
1. How was communist influence visible in Japan immediately after the war? What measures were taken to combat this? How successful were they?

What are the catch-phrases used today by the Japanese Communist Party to arouse feeling?
2. Investigate modern Japan's attitude to China and the United States, keeping in mind what you know of the historical background.
3. Summarise the terms of the San Francisco Treaty (1951). Why was this treaty promulgated at this time? What was the attitude of Russia, Australia and a section of the Japanese people?
4. What are the main problems facing modern Japan?
5. Make sure your time chart is up to date.
6. Why do you think Japan made such a rapid recovery after the war, and how do you account for her present position of economic strength?

Dates of Important Events

60 BC	Jimmu Tenno
536 AD	Rise of Soga family
645	Rise of the Fujiwaras
646	Taika or Great Reform
702	Taiho Laws
710	Foundation of Nara
794	Capital moved to Kyoto
1000	*Tale of Genji*
1160	The Taira defeat the Minamoto in Gempei War
1180	Yoritomo and Yoshitsune take up arms against the Taira
1183	The Taira flee from Kyoto
1185	Battle of Dannoura
1192	Yoritomo becomes Shogun at Kamakura
1274	Mongol Invasion
1281	Second Mongol Invasion
1318	Go-Daigo comes to throne
1331	He takes up arms against the Regent
1333	Defeat of the Hojos
1336	Ashikaga Takauji becomes Shogun
1392	End of the 'false Emperors'
1534	Birth of Oda Nobunaga
1542–3	Portuguese traders visit Japan
1549	St Francis Xavier arrives
1567	Nobunaga enters Kyoto
1582	Death of Nobunaga: Hideyoshi takes over
1587	Jesuits ordered to leave Japan
1590	Victorious Hideyoshi transfers his capital to Tokyo
1591	Hideyoshi claims sovereignty over the Philippines
1592	He attacks Korea

Art Periods

Haniwa period
552 AD

Asuka Period

710
Nara Period

794

Heian Period
(Heian-kyo = Kyoto)

1185

Kamakura Period

1333

Muromachi Period

1568

146 Japan

1597	Execution of Christians: second attack on Korea	
1598	Death of Hideyoshi	
1600	Battle of Sekigahara: Will Adams in Japan	*1600*
1603	Iyeyasu becomes Shogun	
1605	He retires in favour of his son, Hidetada	
1614	Christianity outlawed: Hideyori crushed at Osaka	
1616	Death of Iyeyasu	Edo or
1636	Iyemitsu begins the 'Seclusion Period'	Tokugawa Period
1638	Christianity banned: Massacre of Shimabara	
1716	Embargo on Dutch books lifted by Yoshimune	
1846	Commodore Biddle visits Japan	
1847	Emperor Komei ascends the throne	
1853	Commodore Perry ends Japan's isolation: Sakhalin annexed by Russia	*1853*
1854	Treaty of Kanagawa	Modern Period
1855	Russo-Japanese Treaty	
1856	Townsend Harris arrives in Japan	
1858	Treaty of Commerce and Navigation	
1867	Death of Emperor Komei	
1868	Battle of Toba-Fushimi: Meiji Restoration	
1877	Saigo's rebellion	
1881	Imperial Rescript promises a parliament	
1885	Treaty of Tientsin	
1889	Meiji Constitution handed to Kuroda	
1892	Beginning of Trans-Siberia Railway	
1894	Sino-Japanese War	
1895	Treaty of Shimonoseki 'Triple Intervention': Taiwan annexed by Japan	
1898	Russia acquires lease of Liao-tung Peninsula: Hawaii and the Philippines annexed by the United States	
1900	Boxer Rebellion	
1902	Anglo-Japanese alliance	
1904	Russo-Japanese war	
1905	Treaty of Portsmouth: Japan acquires South Sakhalin	
1910	Korea annexed by Japan	
1914	First World War begins	
1915	The 'Twenty-One Demands'	
1917	Revolution in Russia	
1918	End of First World War	
1921–22	Washington Conference	
1923	Tokyo earthquake	

1926	Emperor Hirohito ascends the throne
1932	Japanese set up puppet government in Manchuria
1933	Lytton commission on China: Japan leaves the League of Nations
1936	Army plot
1937	Prince Konoye becomes Prime Minister
1940	Tripartite Axis Pact
1941	Attack on Pearl Harbour: Japanese troops land in Malaya
1942	Battle of the Coral Sea: Battle of Midway
1944	The United States bombs Japan: Battle of Kohima-Imphal: Battle of Leyte Gulf
1945	Atomic bombs on Hiroshima and Nagasaki: Russia declares war on Japan: The Japanese surrender (September)
1946	Yoshida becomes Prime Minister: new Japanese constitution
1951	San Francisco Peace Treaty: USA-Japanese Security Treaty: Anzus Pact
1960	Assassinations of political leaders: demonstrations against President Eisenhower's visit to Japan
1968	Bonin Islands returned to Japan

Index

Adams, Will, 37, 43, 132-34
Agriculture, 1, 2, 9, 10, 119
Aikokusha, 68, 69
Ainu, 7, 9, 12, 51
Akihito, Crown Prince, 120, 123
Alaska, 52
Alcock, Sir Rutherford, 131
Aleutians, 101
Amami-Oshima, 101
Amaterasu, 8, 58
Amboina, 41
Amida Buddhism, 32, 34
Amur River, 52, 64
Anglo-American Treaty (1914): see Treaties
Anglo-Japanese Alliance (1902), 80, 101
Anti-Comintern Pact (1936), 105
Anzus Pact (1951), 122
Arabs, 39, 40
Araki, General, 96, 97
Architecture, 3, 4
Army of Occupation, 119, 121
Art, 5, 12, 27, 28, 32, 130
Asano, 129
Asanuma, 123
Ashikaga Shogunate, 26-29
Ashikaga, Takauji, 26, 27
Ashikaga Yoshimitsu, 59
Asia, 1, 15, 51, 64, 80, 82, 101, 107, 108, 110, 114, 118, 119, 123, 125
Assam, 111
Atlantic, 132
Australia, 98, 105, 109, 110, 117, 122, 123, 124

Austria, 69

Bakufu, 23
Balkans, 82
Ball, W. McMahon, 119
Baltic, 80
Bank of Japan, 88, 97
Banks, 67, 88, 90, 91
Belgium, 101
Bellinger, Vice-Admiral, 113
Berlin, 105, 107
Biddle, Commodore, 50
Bismarck, 67
Black Dragon Society, 131
Black Sea, 76
Bokhara, 39
Bonin Islands, 5, 76, 101, 121
Borneo, 109
Bose, Subhas Chandra, 111
Boxer Rebellion, 79, 104
Brisbane, 109
British East India Company, 41, 134
Buddha, 11, 12, 31, 56
Buddhism, 11, 12, 13, 21, 29, 31, 33, 34, 35, 58, 59, 66, 71, 124, 130, 135
Burma Road, 104, 107
Buryats, 7
Bushido, 21, 95
Byodoin Temple, Nara, 56

California, 84, 85
California Land Act (1913), 85
Caroline Islands, 101
Cavendish, Thomas, 41

150 Japan

Chang Tso-Lin, 102
Charter Oath, 68
Chiang Kai-shek, 5, 99, 102, 103, 104, 107, 108
Chikamatsu Monzaemon, 60, 61
China, 3, 5, 6, 7, 8, 11, 12, 13, 14, 15, 24, 28, 29, 30, 32, 36, 39, 44, 46, 49, 50, 51, 58, 64, 76, 77-81, 84, 85, 86, 95, 96, 97, 98, 99, 100, 101, 102, 103, 104, 105, 106, 109, 110, 111, 114, 117, 122, 131
Chinese Eastern Railway, 81, 102
Choshu, 34, 35, 56, 67, 71, 72, 73, 74, 95, 131
Christians, 14, 31, 34, 35, 37, 39, 40, 42, 43, 44, 67, 71, 130, 133
Chungking, 104
Churchill, Sir Winston, 112
Chushingura, 61
Climate, 3
Coal, 2, 6, 81, 90, 99, 120
Coinage, 29, 36, 46, 49
Communications, 2, 6,
Communists, 6, 15, 93, 99, 103, 104, 116, 118, 120, 121, 122, 124
Confucius, 14, 30, 32, 72, 126, 130
Constitution of Japan (1946), 117
Coral Sea, Battle of (1942), 109
Crimean War, 52

Daimyo, 21, 22, 28, 29, 37, 41, 42, 43, 47, 48, 49, 54, 55, 63, 64, 68, 73, 87, 88
Dairen, 77, 79
Dan, Baron, 91, 96
Dance, 58, 59, 60
Dannoura, Battle of (1185), 22
Deshima, 44
Drake, Sir Francis, 40
Drama, 58-62
Dutch: *see Holland*
Dutch East India Company, 41
Dutch East Indies: *see Indonesia*

Earthquakes, 1, 3, 90, 94
Edo: *see Tokyo*
Eisenhower, General, 123
England, 37, 40, 41, 50, 51, 52, 53, 56, 57, 64, 67, 69, 70, 77, 78, 81, 82, 84, 86, 100, 104, 105, 106, 107, 109, 110, 111, 112, 117, 122, 123, 132, 133, 134
Europe, 37, 39, 40, 43, 69, 86, 101, 107, 108, 114
Extra-territorial Rights, 52, 53, 67, 78

Far Eastern Commission, 117
Farmers, 2, 28, 29, 35, 36, 47, 48, 68, 91, 119
Fascism, 94, 103, 124
Feudalism, 4, 46, 65, 128-31
Fillmore, President, 50
First World War, 2, 8, 82, 90, 98-100
Fishing, 2
Formosa (Taiwan), 5, 43, 78, 89, 101
Forty-Seven Ronin, 61, 128-31
Four Power Treaty: *see Treaties*
France, 53, 56, 57, 64, 67, 69, 78, 81, 84, 98, 100, 101, 107, 108, 110, 114
Franciscans, 42
French Indo-China, 107, 108, 109, 110, 112, 114, 115
Friendly Society, 93
Fudai, 47
Fuji (Mount), 1
Fujiwara, 12, 13, 14, 20, 21, 22, 23, 35, 63, 97

Gempei War, 22, 59
Genji, Tale of, 13
Genro, 72, 73, 96, 106
Germany, 67, 69, 78, 80, 81, 84, 86, 98, 104, 105, 107, 108, 111, 112
Gneist, Rudolph, 69
Goa, 40, 41
Go-Daigo, Emperor, 26, 27
Go-Komatsu, Emperor, 27
Go-Sanke, 47, 55
Guam, 101, 109
Guilds, 49

Hainan Islands, 105
Hakodate, 50, 53

Index

Hamaguchi Yuko, 94, 95
Hanamichi, 61
Haniwa, 10
Hara Takashi, 94
Harbin, 79
Harris, Townsend, 52
Hashimoto Sanai, 131
Hatoyama Ichiro, 123
Hawaii, 84, 85, 109, 113
Heian-kyo: see *Kyoto*
Heian Period, 13
Hemi, 133, 134
Henry Pu-yi, 103
Hidetada, Shogun, 37, 134
Hideyori, 36, 37, 128
Hideyoshi: see *Toyotomi*
Higashikuni, Prince, 115
Hirado, 41, 133, 134
Hiragana, 16
Hirohito, Emperor, 95, 115, 120
Hirosaki Castle, 57
Hiroshima, 4, 112
Hirota Koki, 118
Hitachi, 89, 120
Hitler, 106, 107, 108
Hizen, 56, 131
Hojo Clan, 23, 24
Hojo Takatoki, 26
Hojo Tokimasa, 22
Hokkaido, 1, 2, 4, 5, 7, 9, 51, 68, 69, 88, 90, 104, 116, 120, 125
Holland, 37, 41, 43, 44, 49, 50, 51, 52, 53, 56, 64, 101, 108, 109, 110, 114, 117, 132, 133, 134
Hong Kong, 89, 101, 109
Honshu, 1, 2, 4, 8, 9, 10, 34, 41, 51
Hupei, 99

Ikeda Hayato, 97, 123
India, 3, 12, 31, 32, 39, 40, 41, 53, 64, 90, 110, 117
Indian Independence League, 110
Indonesia, 9, 40, 41, 64, 108, 109, 115, 126
Industrial Rationalisation Bureau, 91
Inouye Kaoru, 70, 96

Instrument of Surrender, 115
Inukai Tsunayoshi, 96
Iron Ore, 2, 90
Iruka, 12
Itagaki Taisuke, 73
Italy, 101, 107, 111
Ito Hirobumi, 67, 69, 70, 73, 77, 80, 81, 82, 84
Iwojima, 111
Iyemitsu, 44
Iyeyasu: see *Tokugawa Iyeyasu*
Izanagi, 8
Izanami, 8
Izumo, 8, 9, 60

Java, 41
Jesuits, 41, 42, 43
Jimmu Tenno, 8
Jingo, Empress, 11
Jiyuto, 69
Jomon Period, 9
Joruri: see *Minstrels*
Judo, 32

Kabuki, 58, 60-62, 72
Kagoshima, 41, 56
Kagura, 58
Kaishinto, 69
Kaiso Kaisha, 89
'Kakizome', 73
Kamakura, 5, 23, 26, 27
Kamikaze, 24, 111
Kampaku, 13, 35
Kanagawa, 53
Kanagawa, Treaty of: see *Treaties*
Kanji, 16
Kanto, 1, 27, 35
Karafuto: see *Sakhalin*
Katayama Sen, 93
Kawasaki, 89
Kenseihonto, 73
Kenseito, 73, 74
Kira, 129, 130
Kishi Nobusuke, 123
Knox, P. C., 86
Kobe, 4, 53, 57
Kogon, 27

152 Japan

Kohima-Imphal, 111
Kojiki, 12, 55, 58
Komei, Emperor, 53, 57
Komeito, 134
Konoye, Prince, 96-97, 100, 118
Korea, 6, 8, 11, 12, 15, 24, 32, 36, 42, 64, 65, 76-82, 84, 94, 105, 121, 126, 131
Kotoku Denjiro, 93
Kota Bharu, 109
Kublai Khan, 24
Kuomintang, 99
Kurile Islands, 5, 52, 76, 101, 112
Kuroda Kiyotaka, 70
Kwantung Army, 102
Kyogen, 59
Kyoto, 1, 5, 13, 22, 23, 26, 27, 29, 34, 35, 46, 47, 55, 57
Kyushu, 1, 2, 4, 8, 9, 10, 24, 35, 90, 132

Labour Relations Commission, 120
Labour Union Law, 116
Language, 15-19
League of Blood, 96
League of Nations, 103, 106
Leyte Gulf, Battle of, 111
Liaotung Peninsula, 6, 77, 78, 79, 81
Liefde, 132, 133
Li-Ito Convention, 77
London, 80
Loochoo Islands: *see Ryukyu*
Lytton Commission, 103

Macao, 41
MacArthur, General, 115, 116, 117, 120, 121
Magellan, 40
Makiguchi Tsunesaburo, 135
Malacca, 40
Malaya (and Malays), 5, 7, 40, 109, 110, 112, 114
Manchu Dynasty, 79, 98-99, 125, 126
Manchukuo Railway Company, 91
Manchuria, 6, 10, 79, 80, 81, 84, 86, 91, 95, 97, 99, 100, 101, 102, 103, 104, 105, 112

Manchuria Incident, 103
Manyoshu, 12
Marco Polo Bridge Incident, 104
Marianas, 101, 111
Martin, General, 113
Matsukata Masayoshi, 73, 89
Matsuoka Yosuke, 107
Meiji Constitution, 70-75, 95, 115
Meiji, Emperor, 57, 105
Meiji Restoration, 27, 58, 63-69, 73, 128
Merchants, 28, 29, 45, 46, 47, 48, 60, 68, 74, 88
Midway, Battle of (1942), 110
Minamoto Clan, 22, 23
Minseito, 93, 94
Minstrels, 59, 60
Missouri, 115
Mitsubishi, 89, 94
Mitsuhide, 34
Mitsui, 46, 56, 69, 87, 88, 89, 91, 96, 97
Mitsukuni, 55
Miura, Viscount, 78
Mongolia (and Mongols), 6, 24, 84, 103, 104
Moratorium, 48, 90
Moscow, 93, 107
Moscow Conference, 116
Mount Fuji: *see Fuji, Mount*
Mukden, 80, 103
Murasaki Shikibu, Lady, 13
Muraviev, Count, 52
Muromachi Era, 27, 48
Myths, 7

Nagasaki, 4, 42, 43, 44, 49, 50, 51, 52, 53, 112, 132, 133
Nagoya, 111
Naka, Prince, 12, 13
Nakamura, Captain, 102
Nan Han Islands, 77
Nanking, 104
Napoleon III, 57
Nara, 5, 12, 13, 46, 130
National Personnel Authority, 120
New Guinea, 109, 110

New Stone Age, 8
New Year's Day, 73, 121
New Zealand, 117, 122
Nichiren Buddhism, 29, 32, 135
Nihon Shoki (or Nihongi), 12, 55
Niigata, 53
Nikko, 5
Nippon Yusen Kaisha, 89
Nobunaga: see Oda
Noh Plays, 28, 58, 59, 72
Nomon-han, 107
Northern Emperors, 27
Nozaka Sanzo, 120

Oda Nobunaga, 34, 35, 41, 42
Oil, 5, 6, 91, 92, 108, 109, 112, 114, 125
Oishi, 129
Okada, Admiral, 96
Okinawa, 5, 111, 122
Okuma Shigenobu, 70, 72, 73
O-Kuni, 60
Ono Family, 87
Open Door Policy, 85, 100
Opium War, 50
Osaka, 2, 4, 35, 37, 53, 56, 60, 68, 128, 133
Owen Stanley Ranges, 110

Pacific, 40, 41, 51, 52, 84, 98, 101, 110, 132
Pago Pago, 101
Panama Canal, 85
Papua, 109
Parkes, Sir Henry, 56
Peace Preservation Act, 94
Pearl Harbour, 109, 110, 113, 114
Peking, 77, 79, 51, 104
Pelew Islands, 101
Perry, Commodore, 44, 45, 50, 52
Pescadores, 78, 101
Philippines, 9, 40, 41, 42, 84, 101, 109, 111, 117, 126, 133
Political Parties, 68, 69, 72, 73, 74, 91, 96, 120, 121, 134
Pope, 40, 42
Port Arthur, 77, 79, 80

Port Hamilton, 77
Port Lazarev, 77
Port Moresby, 109, 110
Portsmouth, Treaty of: see Treaties
Portuguese, 34, 35, 37, 40, 41, 42, 101, 132, 133
Products, 2, 4
Public Safety Commission, 119
Puppets, 59
Putyatin, Admiral, 52

Railways, 2, 67, 79, 82, 86, 91, 100, 101, 103
Raw Materials, 2, 105, 108
Reinsch, Paul S., 99
Rescript on Education, 72
Ribbentrop, 105
Rice, 2, 10, 29, 45, 46, 48, 75, 90, 91
Richardson, 56, 128
Riots, 28, 29, 48, 90
Roches, Leon, 57
Ronin, 28, 128-31
Roosevelt, Franklin D., 112
Roosevelt, Theodore, 81, 85
Russia, 2, 5, 6, 7, 40, 51, 52, 53, 64, 65, 69, 76-82, 84, 86, 93, 96, 101, 102, 103, 104, 105, 107, 108, 112, 116, 117, 118, 121, 122, 132
Russian Revolution (1917), 5, 93, 100
Russo-German Pact (1939), 108
Russo-Japanese Treaty: see Treaties
Russo-Japanese War (1904-1905), 80-83
Ryukyu Islands, 5, 43, 76, 101, 111, 121

Saigo Takamori, 64-65, 131
Saionji, Prince, 72, 106
Saipan, 111
Saito, Admiral, 96
Sakhalin, 5, 50, 52, 81, 102, 112
Samoyeds, 7
Samurai, 21, 22, 24, 28, 36, 37, 47, 48, 61, 64, 65, 68, 74, 88, 128
San Francisco, 85
San Francisco Peace Treaty: see Treaties

154 Japan

Sapporo, 1, 4, 88, 104
Satsuma, 34, 35, 41, 56, 64, 71, 72, 73, 89, 131
Second World War, 107-14
Seiyukai (Kenseito), 74, 93
Sekigahara, Battle of, 36, 37, 47, 54
Sendai, 43
Sengakuji, 130
Seoul, 77, 81
Shanghai, 89, 104
Shantung, 99, 100, 101
Shidehara, Baron, 116
Shigemitsu Mamoru, 115
Shikoku, 1, 35
Shimabara, 44
Shimada Family, 87
Shimoda, 50, 52
Shimonoseki, Straits of, 22, 56
Shimonoseki, Treaty of: *see Treaties*
Shingon Buddhism, 32
Shinto, 10, 11, 12, 30, 32, 55, 58, 66, 130, 135
Shipbuilding, 2, 4, 92
Shizuoka, 37
Shoen, 20, 45
Shotoku, Prince, 11, 12, 58
Showa Era, 18, 95
Showa Restoration, 95
Siam, 15, 109, 112, 114
Siberia, 9, 52, 81, 100, 102
Silk, 39, 88, 90, 91
Singapore, 101, 107, 109
Sinkiang, 103
Soga Clan, 11, 12
Soka Gakkai, 124, 134-35
Southern Emperors, 27
Spaniards, 37, 40, 41, 42, 43, 84, 132, 133
Spice Islands, 39, 40
Spratley Islands, 105
Stalin, 107, 112
Steel, 2, 4, 92, 99
Suicide, 21, 115, 118, 130
Suinin, Emperor, 10
Sumatra, 40
Sumitomo, 89
Sun Yat-sen, 99

Suzuki, 96
Szechuan, 99

Taiho Laws, 14, 63
Taika Reforms, 13, 45, 63
Taira Clan, 22, 23
Taisho, Emperor, 95
Takachiho, Mount, 8
Tanaka, Baron, 102
Tendai Buddhism, 32
Ternate, 40
Thailand, 32
Tientsin, 105
Tientsin, Treaty of: *see Treaties*
Toba-Fushimi, Battle of, 57, 131
Togo, Admiral, 81
Tojo, General, 109, 118
Tokaido, 133
Tokugawas, 9, 45, 59, 63
Tokugawa Iyeyasu, 34, 35, 36, 37 43, 47, 48, 49, 54, 55, 128, 132
Tokusei, 48
Tokyo (Edo or Yedo), 1, 2, 4, 8, 9, 10, 27, 44, 47, 53, 57, 63, 67, 73, 81, 90, 93, 94, 96, 111, 112, 115, 117, 118, 120, 130
Tosa, 56, 131
Toyotomi Hideyoshi, 34, 35, 36, 37, 42, 43, 132, 133
Tozama, 47
Trade Treaty: *see Treaties*
Trans-Siberian Railway, 77, 79
Treaties:
  Anglo-American (1914), 86
  Commerce and Navigation (1858), 53
  Four Power (1921-22), 101
  Kanagawa (1854), 50, 52
  Portsmouth (1905), 81
  Russo-Japanese (1855), 52
  San Francisco (1951), 121
  Shimonoseki (1895), 77, 79
  Tientsin (1885), 77
  Trade Treaty (1911), 105
  US-Japanese Security Treaty (1951), 122, 123
Tripartite Axis Pact (1940), 107

Triple Intervention, 78
Tsinan-Tsingtao Railway, 101
Tsingtao, 98
Tsushima, 76, 81
Twenty-One Demands, 99

Ugaki, General, 97
United States of America, 2, 5, 6, 44, 50, 51, 53, 56, 57, 80, 84, 85, 86, 91, 100, 101, 104, 105, 106, 108, 109, 110, 111, 112, 113, 114, 115, 117, 118, 121, 122, 123, 130
Uraga, 43
US-Japanese Security Treaty (1951): see Treaties
Uzume, 58

Vassilievsky, Marshal, 116
Versailles Conference, 100
Vining, Mrs, 123
Vladivostok, 64, 77, 79, 89
Voluntary Loans, 29, 49

Wake Island, 109
Washington Conference (1921-22), 79, 101-02
Weihaiwei, 77

Xavier, Saint Francis, 41

Yalta Conference, 112
Yamagata Aritomo, 73, 74, 81, 95
Yamaguchi, 42
Yamamoto, Admiral, 113
Yamato, 8, 10
Yangtze River, 104
Yasuda Zenjiro, 89
Yayoi Culture, 9
Yedo: see *Tokyo*
Yezo, 1, 5
Yodo River, 4
Yokohama, 4, 52
Yokohama Specie Bank, 88
Yokosuka, 57, 133
Yoritomo, 22, 23, 27
Yoshida Shigeru, 116, 120, 121, 123
Yoshida Shoin, 131
Yoshimune, 49
Yoshino, 27
Yoshitsune, 22, 23
Young, Marshal, 102, 103

Zaibatsu, 87, 89, 90, 91, 92, 93, 96, 97, 116, 118, 122
Zen Buddhism, 32-33